"As a statistics teacher, I'm always on the lookout for accessible yet challenging problems for students to chew on in class. The problems in this book are wonderful: they're clear, they test important concepts in statistics, and they require insight without being tricky."
— **Andrew Gelman**, *Professor of Statistics and Political Science, Columbia University, Author of Active Statistics: Stories, Games, Problems*

"Writing great test questions can among the most challenging aspects of teaching. We draw on Smith's collection extensively in our "Calling Bullshit" course on data reasoning. Perhaps I should not disclose the source of my exam questions — but if any of my students choose to seek an advantage by reading Smith's book from cover to cover, they'll learn more by doing so than I could hope to teach them in a quarter."
— **Carl T. Bergstrom**, *Professor of Biology, University of Washington, Author of Calling Bullshit: The Art of Skepticism in a Data-Driven World*

"A treasure-trove; a gold mine! Buy this book if you want questions that require critical thinking (rather than multiple-choice guessing). Hundreds of questions and answers carefully organized by topic. Every one teaching introductory statistics should own a copy of this book!"
— **Milo Shield**, *American Statistical Association Fellow, Visiting Professor, New College of Florida, Author of Statistical Literacy and Editor of www.StatLit.org*

"Most professors can write tests that measure knowledge, but Gary Smith's questions often go well beyond this to measure a student's ability to use this knowledge. What a remarkable collection of superb thought-provoking questions!"
— **Woody Studenmund**, *Professor of Economics, Occidental College, Author of Using Econometrics*

"The unique features of these exam questions are their origins in everyday life and insights into ferreting whether students know the subject and can apply it in nonrobotic ways. I wish I had access to this test bank when I began teaching."
— **William E. Becker**, *Professor Emeritus of Economics, Indiana University, Past Editor, The Journal of Economics Education*

"Professor Smith's questions draw upon interesting real-world scenarios to test students' understanding of key statistical concepts. I wish I had had this collection when I first started teaching statistics!"
— **Alan Reifman**, *Professor of Human Development & Family Sciences, Texas Tech University, Author of Hot Hand: The Statistics Behind Sports' Greatest Streaks*

Exercises and Solutions in Probability and Statistics

The book contains hundreds of engaging, class-tested statistics exercises (and detailed solutions) that test students' understanding of the material. Many are educational in their own right—for example, baseball managers who played professional ball were often catchers; stocks that are deleted from the Dow Jones Industrial Average have generally done better than the stocks that replaced them; athletes may not get hot hands but they often get warm hands with modest improvements in their success probabilities.

Gary Smith is the Fletcher Jones Professor of Economics at Pomona College. He has won two teaching awards and written (or co-authored) more than 100 academic papers and 20 books. His statistical and financial research has been featured in various media, including *The New York Times*, *Wall Street Journal*, *Wired*, *NPR Tech Nation*, NBC Bay Area, CNBC, WYNC, WBBR Bloomberg Radio, *NBC Think*, *Silicon Valley Insider*, *Motley Fool*, *Scientific American*, *Forbes*, *MarketWatch*, MoneyCentral.msn, *NewsWeek*, *Fast Company*, *The Economist*, *MindMatters*, *OZY*, *Slate*, and *BusinessWeek*.

Texts in Statistical Science Series

Joseph K. Blitzstein, Harvard University, USA
Julian J. Faraway, University of Bath, UK
Martin Tanner, Northwestern University, USA
Jim Zidek, University of British Columbia, Canada

Recently Published Titles

Fundamentals of Mathematical Statistics
Steffen Lauritzen

Modelling Survival Data in Medical Research, Fourth Edition
David Collett

Applied Categorical and Count Data Analysis, Second Edition
Wan Tang, Hua He, and Xin M. Tu

Geographic Data Science with Python
Sergio Rey, Dani Arribas-Bel, and Levi John Wolf

Models for Multi-State Survival Data
Rates, Risks, and Pseudo-Values
Per Kragh Andersen and Henrik Ravn

Spatio–Temporal Methods in Environmental Epidemiology with R, Second Edition
Gavin Shaddick, James V. Zidek, and Alex Schmidt

A Course in the Large Sample Theory of Statistical Inference
W. Jackson Hall and David Oakes

Statistical Inference, Second Edition
George Casella and Roger Berger

Nonparametric Statistical Methods Using R, Second Edition
John Kloke and Joesph McKean

Generalized Linear Mixed Models
Modern Concepts, Methods and Applications, Second Edition
Walter W. Stroup, Marina Ptukhina, and Julie Garai

Analysis of Categorical Data with R, Second Edition
Christopher R. Bilder, Thomas M. Loughin

Applied Nonparametric Statistical Methods, Fifth Edition
Nigel C. Smeeton, Neil H. Spencer, and Peter Sprent

Linear Models with R, Third Edition
Julian J. Faraway

For more information about this series, please visit: https://www.routledge.com/ Chapman--HallCRC-Texts-in-Statistical-Science/book-series/CHTEXSTASCI

Exercises and Solutions in Probability and Statistics

Gary Smith

CRC Press
Taylor & Francis Group
Boca Raton London New York

CRC Press is an imprint of the
Taylor & Francis Group, an **informa** business

A CHAPMAN & HALL BOOK

Designed cover image: Gary Smith

First edition published 2026
by CRC Press
2385 NW Executive Center Drive, Suite 320, Boca Raton FL 33431

and by CRC Press
4 Park Square, Milton Park, Abingdon, Oxon, OX14 4RN

CRC Press is an imprint of Taylor & Francis Group, LLC

© 2026 Gary Smith

ISBN: 978-1-041-04834-3 (hbk)
ISBN: 978-1-041-04833-6 (pbk)
ISBN: 978-1-003-63015-9 (ebk)

DOI: 10.1201/9781003630159

Typeset in Palatino
by KnowledgeWorks Global Ltd.

Contents

Introduction ix

1. Mean, Median, and Descriptive Statistics 1

2. Graphs: Good, Bad, and Ugly... 11

3. Misleading Data..36

4. Probabilities.. 52

5. Bayes' Rule 78

6. Monty Hall Problems.. 93

7. Binomial Distribution 99

8. Law of Averages 113

9. Normal Distribution 119

10. One-Sample Tests and Confidence Intervals.. 130

11. Two-Sample Tests and Confidence Intervals.. 148

12. Chi-Square Tests... 170

13. Simple Regression 189

14. Regression toward the Mean..205

15. Multiple Regression 217

16. Miscellaneous.. 241

17. Out-of-Class Projects 266

Index... 289

Introduction

This book contains hundreds of probability and statistics questions (and solutions) that I have used on midterm and final examination tests in the introductory statistics classes that I taught for more than 50 years.

Every question has been class-tested. Questions that did not work well or do not seem particularly interesting have been omitted. My intention here is to share a large set of engaging questions, many of which are educational in their own right—for example, stocks that are deleted from the Dow have generally done better than the stocks that replaced them; lie detector tests are unreliable; athletes may not get hot hands but they often get warm hands with modest improvements in their success probabilities.

In recent years, I have increasingly been teaching interactive statistics classes in which randomly chosen two-person teams work on a projects outside of class and give presentations of their conclusions in class. So, I've also included some suggestions for out-of-class projects in a special Chapter 17.

If you find any typos, fuzzy thinking, or other errors, please let me know so that I can correct them: garynsmith47@gmail.com.

1

Mean, Median, and Descriptive Statistics

1.1 For children under the age of 13, majors is the highest level in Little League Baseball and minors is the next highest level. In the 2023 season, Middletown Little League had enough players to form six majors teams and six minors teams. Instead, they decided to have seven majors teams and five minors teams. What effect did this have on the average quality of the players in the majors and in the minors?

1.2 A college reported that its average class size is 14 students per class. However, a survey that asked this college's students the size of the classes they were enrolled in found that the average class size was 38 students and that 70% of the students' classes had more than 14 students. Provide a statistical explanation for this disparity.

1.3 A school estimated the average number of school-age children per family by questioning each child in the school. Why was their estimate much too high? Give an example to illustrate your reasoning.

1.4 Explain comedian Will Rogers' joke that, "When the Okies left Oklahoma and moved to California, they raised the average intelligence level in both states."

1.5 A sports columnist wrote that "Jon Gruden, left [Monday Night Football] to become Oakland's head coach, a move that has been bad for both the broadcast and the team." Is it possible for someone who goes from one organization to another to lower the average quality of both organizations?

1.6 A survey found that between 1959 and 1983, the average number of hours that women worked outside the home nearly doubled while the number of hours worked inside the home decreased by only 14%. Do these data show that the total number of hours women worked, outside and inside the home, increased between 1959 and 1983?

1.7 A college offers a free book on the history of the college to each graduate whose gift exceeds the median size of the gifts made in the previous year by members of this person's class. Explain why this college either increased or decreased the threshold for a free book by using the median rather than the mean.

DOI: 10.1201/9781003630159-1

1.8 Identify the apparent statistical mistake in this commentary:

> *The median price of a house in Duarte is a whopping $4,276,462, making it the most expensive housing market in the country.... Only 12 homes are currently on the market. So a single high-priced listing (like the mammoth nine-bedroom that's selling for $19.8 million) is enough to skew the median price skyward.*

1.9 A Prudential Financial advertisement titled, "Let's get ready for a longer retirement," said that

> *A typical American city. 400 people. And a fascinating experiment. We asked everyday people to show us the age of the oldest person they know by placing a sticker on our chart.*

Most of the stickers were between the ages of 90 and 110, which Prudential called, "Living proof that we are living longer. Which means we need more money to live in retirement." Why is this conclusion misleading?

1.10 A study divided occupations into five categories and found that women had higher unemployment rates than men in each of the individual categories but, overall, had a lower unemployment rate than men. How is this possible? Use hypothetical data with just two categories to illustrate your reasoning.

1.11 The overall unemployment rate of 10.2% in October 2009 was lower than the 10.8% unemployment rate in November 1982, which was the peak of the worst recession since the Great Depression. However, when the labor force was divided into four educational categories (college graduate, some college education, high-school graduate, and other), the unemployment rate in each category was higher in 2009 than in 1982. How is this possible? Use hypothetical data with just two educational categories to illustrate your reasoning.

1.12 A medical study found that Treatment A was more successful than Treatment B in treating small kidney stones (90% versus 80%) and large kidney stones (70% versus 60%). Is it possible for Treatment B to be more successful overall, when the data for large and small kidney stones were combined? Explain your reasoning and use a numerical example to illustrate your argument.

1.13 A study of the hot hands in basketball included pairs of free throws shot by two players, A and B. When Player A made the first shot, he made the second shot 88% of the time; when he missed the first shot, he made the second shot 91% of the time. Player B was also more likely to make the second shot after missing the first shot: 61% versus 59%. Yet, when the data were combined, the players made 81% of their second shots after making the first shot and only 73% after missing the first shot.

a. How would you explain this reversal? Use hypothetical data on the number of shots each player took to illustrate your reasoning.

b. Which player do you think took more shots?

1.14 Which of these data series has the larger standard deviation?

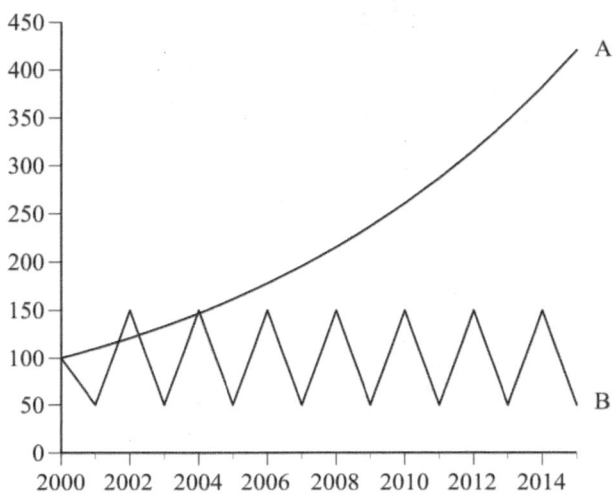

1.15 A researcher found that in the 2009–2010 Premier League season, the soccer teams with the best records tended to win more of their home games than did teams with worse records. He concluded that this demonstrated the importance of a home-field advantage. Why do his data not justify his conclusion?

1.16 When it was proposed that Major League Baseball (MLB) institute a pitch clock that allows a maximum of 20 seconds between pitches, a respected sports columnist wrote that, "If MLB were to institute a 20-second pitch clock and hold players to it, that would, in theory, at this year's pace and pitches per game, have shaved 19 minutes, 48 seconds off the average length of a game." He multiplied 4 seconds (the difference between the 24-second average time between pitches in 2018 and the 20-second pitch clock) times the 297 (the average number of pitches per game): $(24 - 20)\ 297 = 1{,}188$ seconds, which is 19 minutes and 48 seconds. Why is this estimate almost surely wrong? Is the correct time saving most likely larger or smaller than 19 minutes and 48 seconds?

1.17 A study of the effect of college education on income used 1990 and 2010 US Census data. The study looked at: (a) women in the 1990 Census who were between 28 and 32 years old and were also the first female in their family to attend college; and (b) women in the 2010 Census who were between 28 and 32 years old and whose mother was the first female in their family to attend college. The average income was $45,000 for the women in group (a) and $40,000 for the women in group (b), even though average income for the nation as a whole was 50% higher in 2010 than in 1990.

Is it mathematically possible that every daughter had a higher income than her mother, yet the average daughter income was lower than the average mother income? Give a specific numerical example to illustrate your argument.

1.18 A Temple University mathematics professor used these data to show that most Americans have an exaggerated fear of terrorists:

> *Without some feel for probability, car accidents appear to be a relatively minor problem of local travel while being killed by terrorists looms as a major risk of international travel. While 28 million Americans traveled abroad in 1985, 39 Americans were killed by terrorists that year, a bad year—1 chance in 700,000. Compare that with the annual rates for other modes of travel within the United States—1 chance in 96,000 of dying in a bicycle crash, 1 chance in 37,000 of drowning, and 1 chance in only 5,300 of dying in an automobile accident.*

How do you suppose the author calculated the probabilities of dying in a bicycle accident, of drowning, and of dying in a car accident? Do these calculations prove that it is more dangerous to drive to school than to fly to Paris?

1.19 Cameron is taking a class where the final grade is an average of the homework and test scores

$$G = .5H + .5T$$

where H and T are both graded on a scale of 0 to 100. Part way through the semester, Cameron's G was 92. Then Cameron got a score of 90 on a test and the teacher raised Cameron's G to 93.5. Is this possible or did the teacher make a mistake? Use an example to explain your reasoning.

1.20 It has been estimated that as many as 50% of the people infected with COVID-19 were asymptomatic and did not know that they had been infected. If the data on the number of deaths and the number of people who test positive are accurate, but the number of asymptomatic infections is higher than previously estimated, which of the following will be higher, lower, or the same as previously estimated?

a. The percentage of the population that has been infected

b. The percentage of the population that has died from the disease

c. The percentage of the people who were infected that died

d. The percentage of the people who tested positive that died

1.21 Answer this Car Talk puzzler:

> *The preferred footwear in Townberg is combat boots. Twenty percent of the people have one foot, so they wear one boot. Of the remaining people, half go barefoot and half wear two boots—and there are 20,000 boots worn in Townberg. What's the population of Townberg?*

1.22 Market Force asked 7,600 adult Americans how likely they would refer the quick-service pizza restaurant they had most recently visited to others. The results

Blaze	69%
Papa Murphy's	66%
Marco's Pizza	64%
Domino's	48%
Papa John's	47%
Pizza Hut	47%
Little Caesars	45%

As a statistician, how would you criticize a Yahoo news report that described Blaze pizza as "America's favorite pizza chain"?

1.23 A 1994 article in the *American Scientist* reported the results of a survey of 17 scientific experts about the chances of the earth experiencing catastrophic global warming. Here is a summary of their assessments of the probability of a 3°C increase in the global average temperature by 2090. How has this graph been misdrawn?

range = 0% to 30%

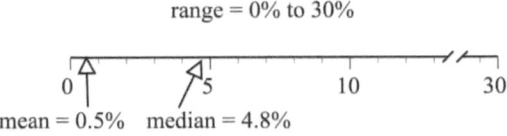

mean = 0.5%   median = 4.8%

1.24 Suppose that you have a sample of ten observations that has a mean of seven, a median of six, and a standard deviation of three. If you subtract two from the value of each observation, what are the new values of the following?

a. Mean

b. Median

c. Standard deviation

If you instead multiply the original value of each observation by two, what are the new values of the following?

d. Mean

e. Median

f. Standard deviation

1.25 Are great discoveries typically made by people who are young and vigorous or old and wise? Below are the ages at which 12 scientists made great discoveries. Use these data to calculate the mean and median ages at which these great discoveries were made.

Scientist	Discovery	Age
Copernicus	Earth revolves around the sun	40
Galileo	Laws of astronomy	34
Newton	Motion, gravitation, calculus	23
Franklin	Nature of electricity	40
Lavoisier	Burning as oxidation	31
Lyell	Earth evolved gradually	33
Darwin	Natural selection in evolution	49
Maxwell	Equations for light	33
Curie	Radioactivity	34
Planck	Quantum theory	43
Einstein	Special relativity	26
Schrodinger	Equations for quantum theory	39

Answers

1.1 It weakened both the majors and the minors. By moving players who would have been above-average in minors and below-average in majors from the minors to the majors, they lowered the average quality of both divisions.

1.2 If we average the class size over the number of classes, a class with 100 students counts as 1 observation and a class with 1 student counts as 1 observation, giving an average class size of $(100 + 1)/2 = 50.5$. If we average the class size over the number of students, a class with 100 students counts as 100 observations and a class with 1 student counts as 1 observation, giving an average class size of $(100 (100) + 1)/101 = 99.0$. (So, 100 out of 101 students are in a class with 100 students.) From the professor's perspective, the average class has 50.5 students. From the student's perspective, the average class has 99 students.

1.3 Families with a large number of children will be overrepresented. Suppose that there are two families, one with one child and one with five children. The average family has $(1 + 5)/2 = 3$ children. If they ask every child, they will get six responses and calculate the average to be $(1 + 5 + 5 + 5 + 5 + 5)/6 = 4.33$.

1.4 The ones who moved must have been below-average in Oklahoma and above-average in California.

1.5 Yes, if Gruden was above average at Monday Night Football and below average as a coach.

1.6 No, because 14% of a large number may be bigger than 100% of a small number. Suppose a person works 10 hours inside the home and 1 hour outside the home every day, a total of 11 hours. If hours inside the home decrease by 14% to 8.6 and hours outside the home double to 2, total hours go down from 11 to 10.6.

1.7 The distribution of gifts is no doubt asymmetrical because gifts cannot be less than zero but can be very large. Those gifts that are very large pull the mean above the median. Thus, the college's use of the median reduced the threshold for a free book.

1.8 An outlier would skew the mean price upward and have little or no effect on the median price.

1.9 These are the ages of the oldest person they know, not a typical person, let alone themselves. Plus, especially in a small town, they may know similar people. Imagine that one person is 110 and everyone in the town knows her. There would be 400 stickers at 110. Finally, the ad doesn't address the question of living longer at all because there are no data for earlier years.

1.10 This is an example of Simpson's paradox. There were more women in occupations that had low unemployment rates.

1.11 This is an example of Simpson's Paradox. Unemployment rates are lower among more highly educated workers, and there were more highly educated workers in 2009 than in 1982.

1.12 This is an example of Simpson's Paradox. The actual data:

	Treatment A		Treatment B	
	Treated	Successes	Treated	Successes
Small stones	87	81 (93%)	270	234 (87%)
Large stones	263	192 (73%)	80	55 (69%)
Combined	350	273 (78%)	350	289 (83%)

Treatment B had a lower success rate for both large and small kidney stones, but had a higher overall success rate. The explanation is that Treatment A was more often used for large stones (which have a relatively low success rate), while Treatment B was more often used for small stones (which have a relatively high success rate).

1.13 This is a real example of Simpson's Paradox (from Robert L. Wardrop). Player A took more shots, with his 88.1% double-hit probability pulling up the overall double-hit probability:

Larry Bird	Second Shot		
First Shot	Hit	Miss	Total
Hit	251	34	285
Miss	48	5	53
Total	299	39	338
Rick Robey	Second Shot		
First Shot	Hit	Miss	Total
Hit	54	37	91
Miss	49	31	80
Total	103	68	171
Bird + Robey	Second Shot		
First Shot	Hit	Miss	Total
Hit	305	71	376
Miss	97	36	133
Total	402	107	509

1.14 B has more volatility but A has the larger standard deviation, which measures the average difference of the observations from the mean value:

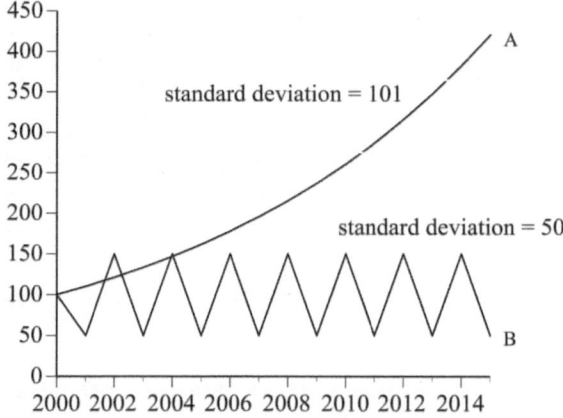

1.15 The teams with the best records may have won (and in fact did win) more of their away games than did teams with worse records. He didn't really measure a home-field advantage because he didn't compare how teams did at home with how they did on road.

1.16   The assertion that the average time savings will be 4 seconds assumes that with a 20-second pitch clock, the average time between pitches will be 20 seconds. However, there will often be less than 20 seconds between pitches; so, the average time between pitches will be less than 20 seconds and the average time saving will be more than 4 seconds. Therefore, the time saving will be more than 19 minutes and 48 seconds.

1.17   This could happen if low-income women have more children than high-income women. Consider four 1990 women, three earning $20,000 and one earning $120,000. Average income is $45,000. Now suppose the low-income women have a total of 14 daughters (hypothetically), each earning $30,000 and the high-income woman has one daughter, who earns $180,000. The average income is $40,000. Even though each daughter earns 50% more than her mother, the average income of the daughters is lower than the average income of the mothers.

1.18   This professor apparently divided the number of Americans killed by terrorists by the number who traveled abroad, divided the number killed in automobile accidents by the total number of Americans, and so on. These are not adjusted for the fact that people may have traveled abroad for only two weeks and driven automobiles for 52 weeks. Also, many trips abroad may have been to places that aren't very dangerous.

1.19   This is possible. If the 90 test score raises the average test score T, it will raise the overall average G. Suppose, for example, that the homework average is H = 100 and there has been one test with a score of T = 84, then G is 92. A score of 90 on the second test will raise the average test score to T = 87 and hence the grade to G = 93.5.

1.20   It is helpful to use hypothetical numbers, such as a population of 1 million of which 100,000 have been infected and 1,000 have died. Now increase the number of infected people from 100,000 to 200,000 and see how this change affects the various statistics.

   a. Higher

   b. Unchanged

   c. Lower

   d. Unchanged

1.21   Twenty percent wear one boot, and the remaining 80% average one boot. So, the average number of boots per person is one. The population of Townberg is 20,000.

1.22   The word "favorite" suggests that this chain sells the most pizzas. Referrals might measure the satisfaction of current customers but be largely unrelated to the number of customers. People who drive Porsches may be very happy with their cars, but Porsches are not America's favorite car.

1.23 The mean should be pulled above the median by the 30 response; the graph, however, is drawn with the mean below the median. The labels are reversed, in that the mean is higher than the median.

1.24 Remember that the mean is the average value, the median is the middle value, and the standard deviation measures the dispersion in the data.

   a. 5
   b. 4
   c. 3
   d. 14
   e. 12
   f. 6

1.25 The mean is 35.42 years; the median is 34 years.

2

Graphs: Good, Bad, and Ugly

2.1 What is misleading about this graph of Affordable Care Enrollment created by a "trusted news source"?

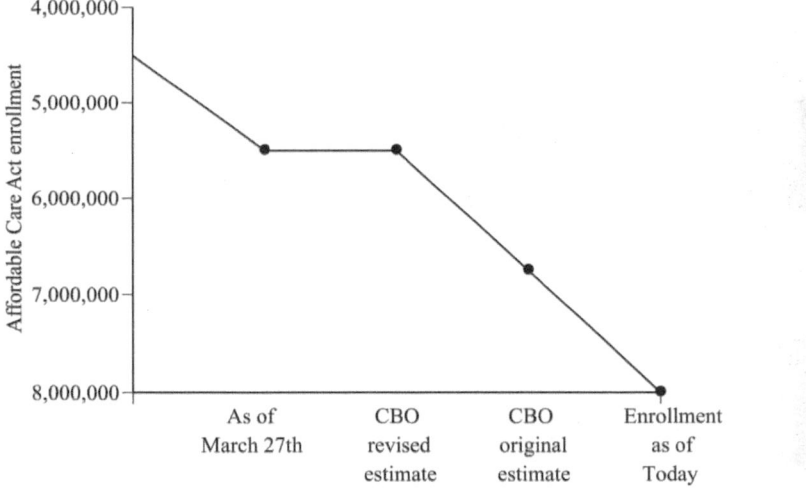

2.2 This Census Bureau figure shows a leveling off of state and local property taxes. How would you improve this figure?

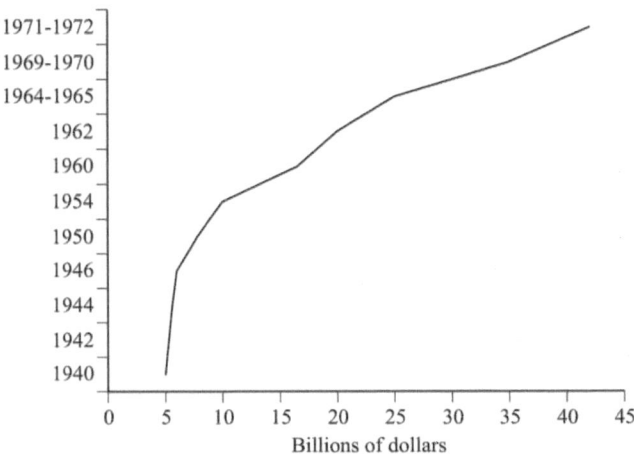

DOI: 10.1201/9781003630159-2

2.3 How would you improve this figure showing the number of Florida murders using firearms before and after the passage of Florida's "Stand Your Ground" law?

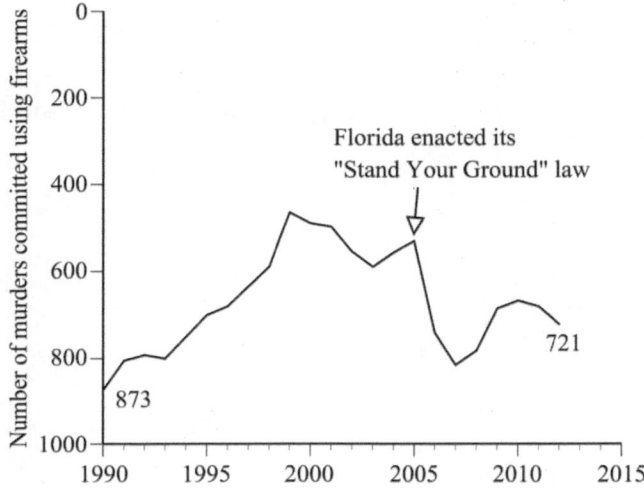

2.4 What is misleading about this figure accompanying a *Los Angeles Times* story that reported, "After peaking at $392 billion in March, money market fund assets have fallen sharply"? How would the visual impression change if this misleading feature was corrected?

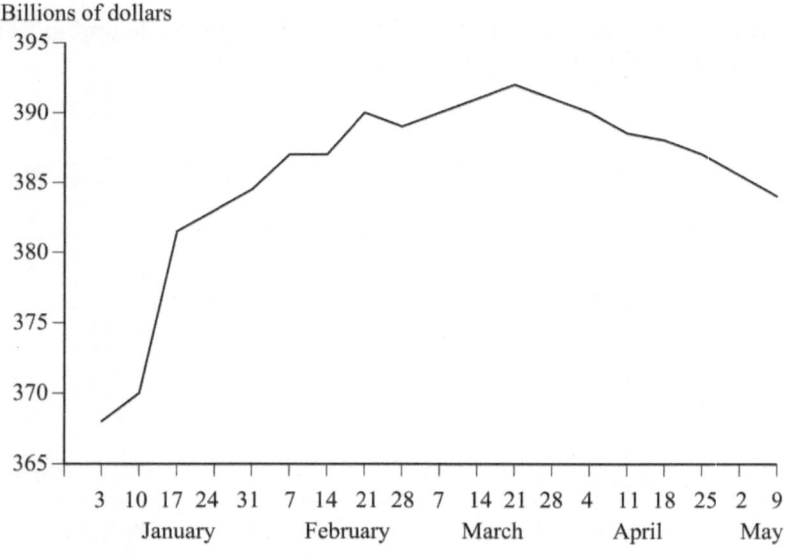

2.5 What is misleading about this figure that Ronald Reagan used to show how much middle-class Americans would save in taxes with the Republican proposal?

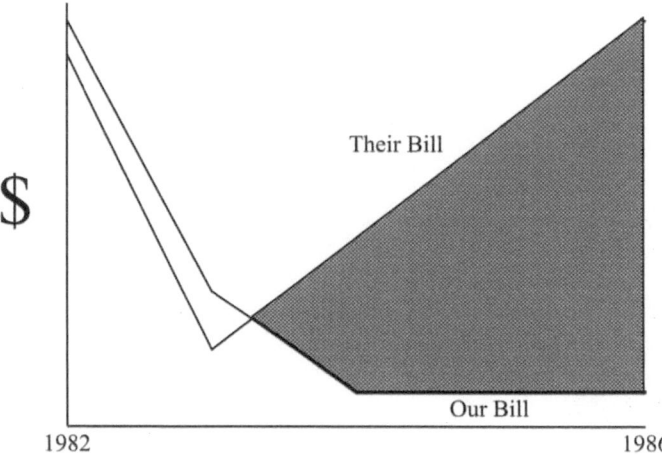

2.6 What is misleading about this National Science Foundation (NSF) figure showing a decline in Nobel Prizes awarded to Americans?

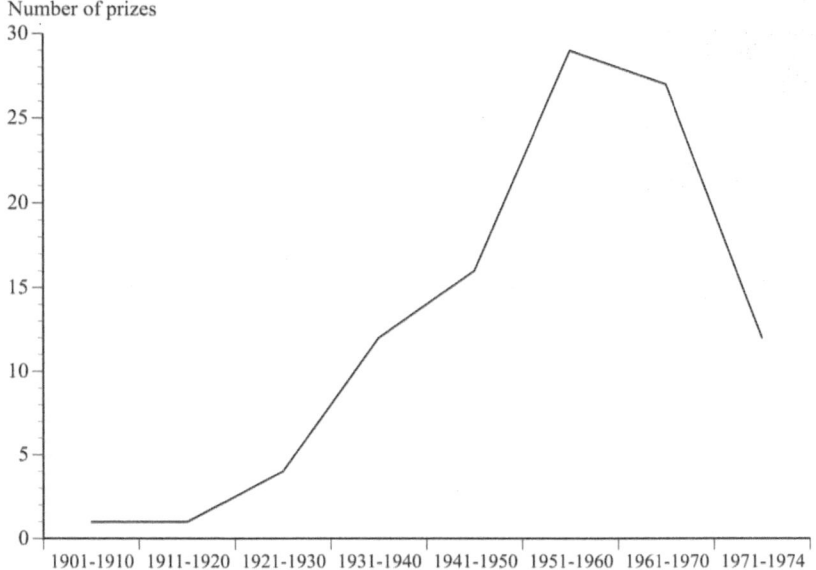

2.7 How would you improve this *Washington Post* figure showing median doctor income over time?

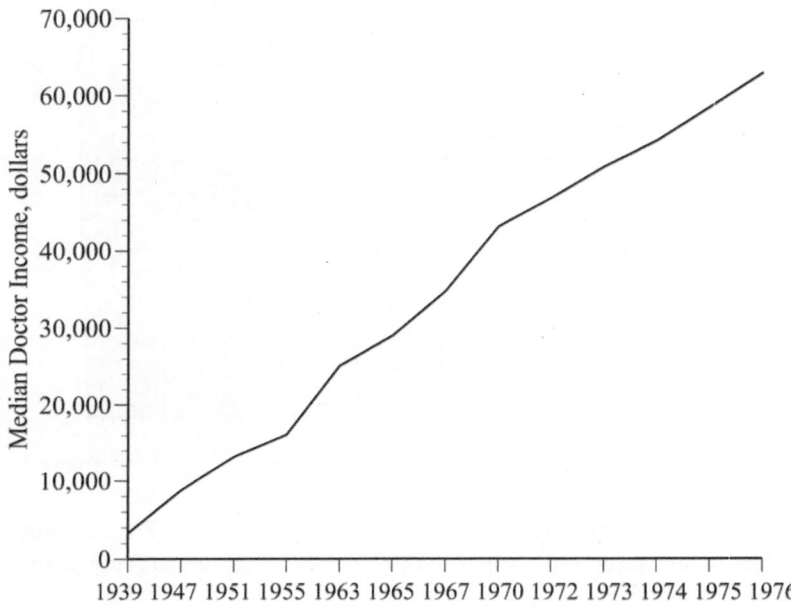

2.8 What do you find misleading about this graph, published on the front page of the *Ithaca Times* on December 7, 2000, evidently showing that Cornell students were paying more and getting less?

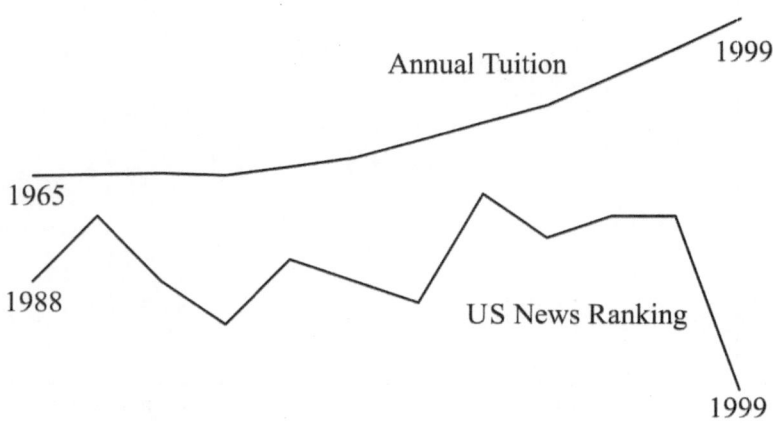

2.9 Explain why this *Los Angeles Times* graph of the Dow Jones Industrial Average of stock prices on January 27, 1997, gives a misleading visual impression of the volatility of stock prices that day:

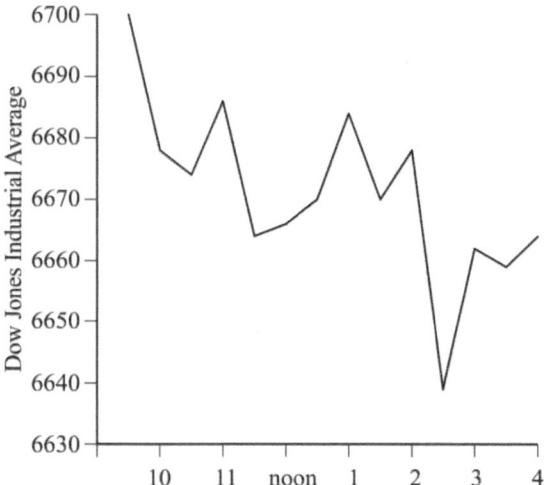

2.10 The April 4, 2016, issue of *Time* magazine contained a full-page advertisement touting gold as an investment: "If you would have taken $150,000 of your money and bought gold on September 6, 2001, then that initial purchase would have been worth over $1 million exactly 10 years later on September 6, 2011!" As a statistician, what problem do you see with this evidence?

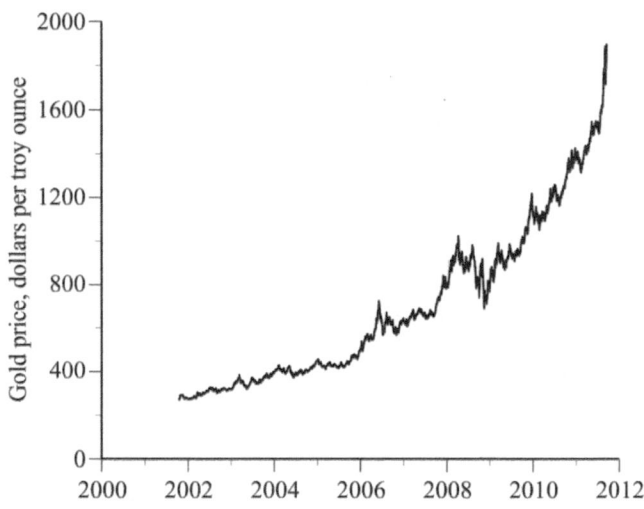

2.11 How would you redo this newspaper figure showing the US income distribution? Don't redraw it, just explain. Do you agree with their definition of the "middle class"?

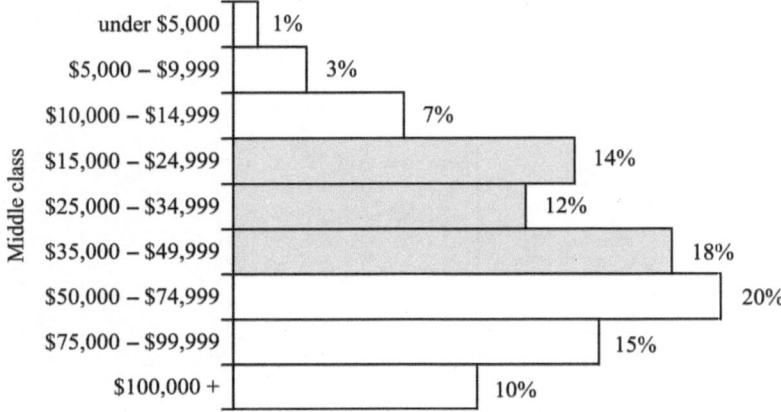

2.12 This *New York Times* figure accompanied a David Frum article titled "Welcome, Nouveaux Riches." The figure shows a dramatic acceleration between 1980 and 1990 in the number of households earning more than $100,000 a year. Frum wrote that, "Nothing like this immense crowd of wealthy people has been seen in the history of the planet." What problems do you see with this graph?

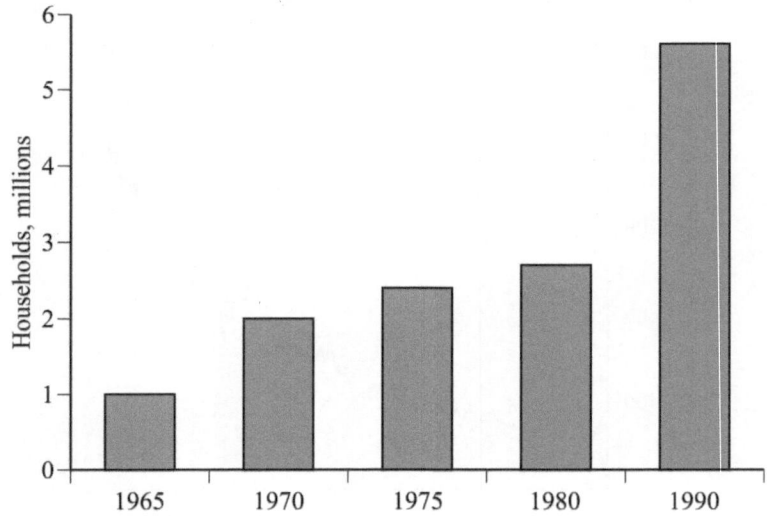

2.13 Identify as many problems as you can find with this figure from a *Forbes* article titled, "True Fact: The Lack of Pirates Is Causing Global Warming."

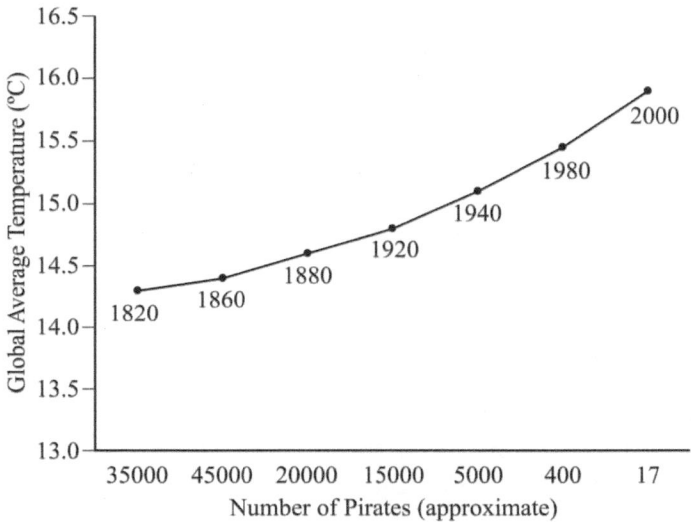

2.14 This figure from *USA Today* shows the eight increases in the cost of a first-class stamp between 1971 and 1991. Identify two distinct reasons why it is misleading.

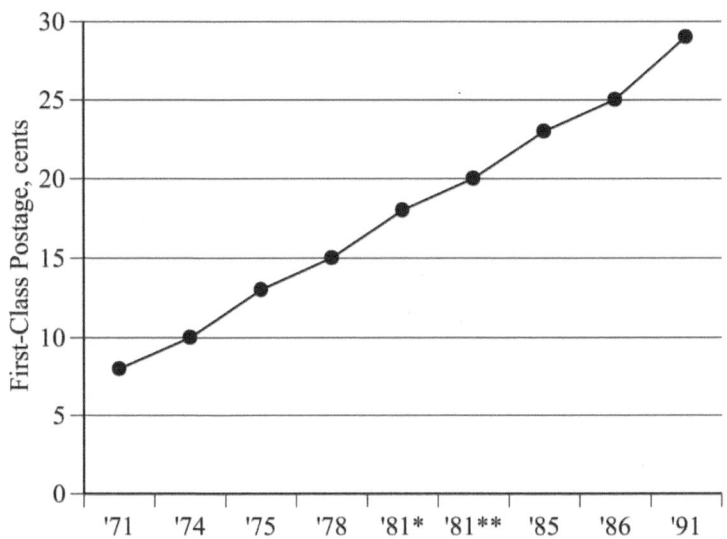

2.15   How was the creator of this figure able to make it seem that Internet Explorer's declining market share had reduced the number of murders in the United States during this 5-year period? What could the creator have done differently to make it seem that they were essentially unrelated?

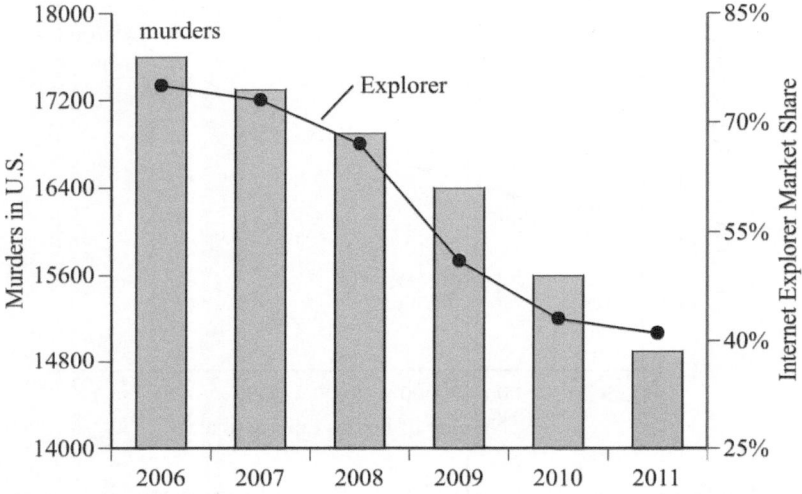

2.16   Why, despite being labeled a histogram, is the following figure not a histogram?

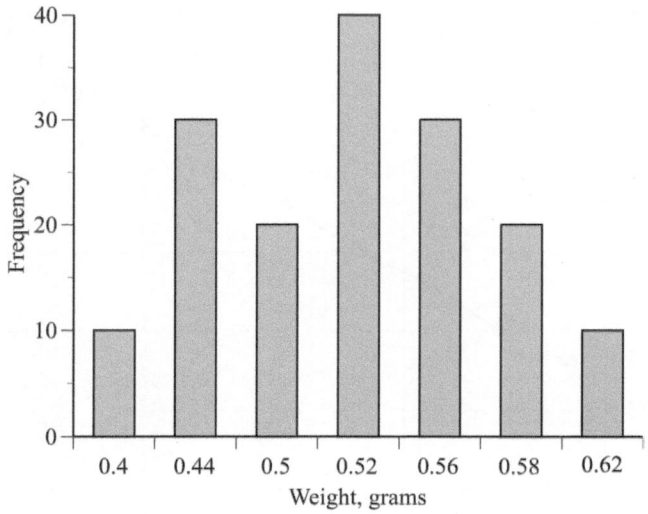

2.17 This figure was used to demonstrate that when the number of parameters in a large language model increases, there is a tipping point beyond which there is an explosive improvement in the model's performance. What unusual feature of this figure may cause the visual impression to be misleading?

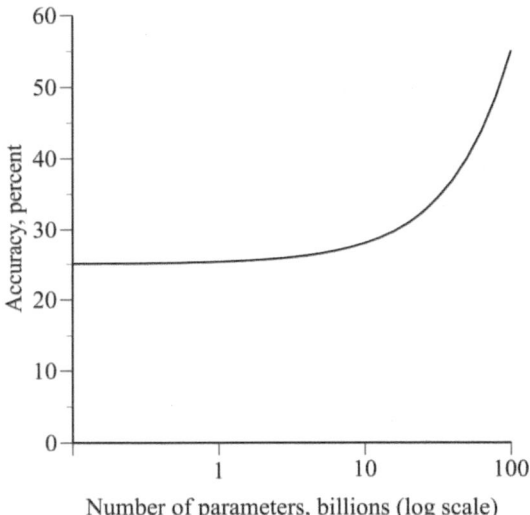

Number of parameters, billions (log scale)

2.18 An economist estimated the "location value" of 12 Native American gambling casinos in Southern California based on the driving distance for gambling households and the competition with other casinos. What problems do you see with this presentation of the results? How would you improve the presentation?

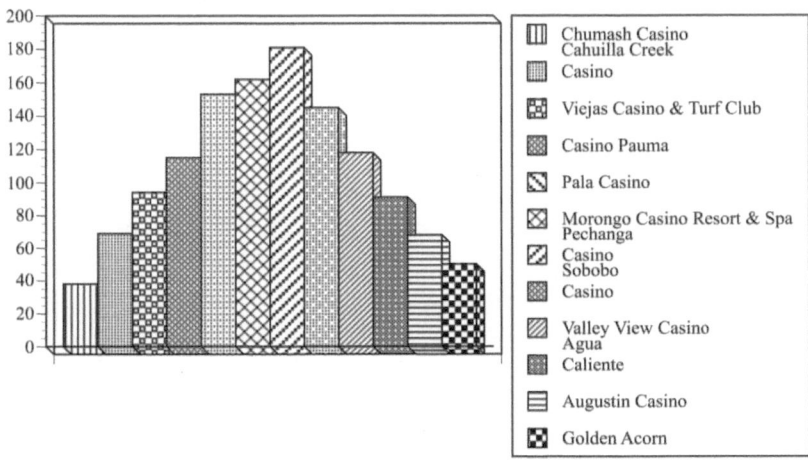

2.19 Historical data on the price of a box seat at Yankee Stadium were used to create this figure, which shows that the increase in ticket prices slowed down during the period of 1995–2010. What would you, as a statistician, say?

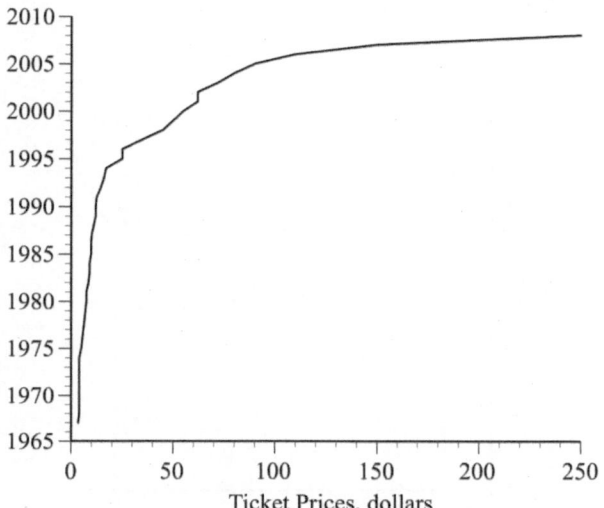

2.20 What is misleading about this figure that was used to show that real (inflation-adjusted) home prices are rising faster than real building costs?

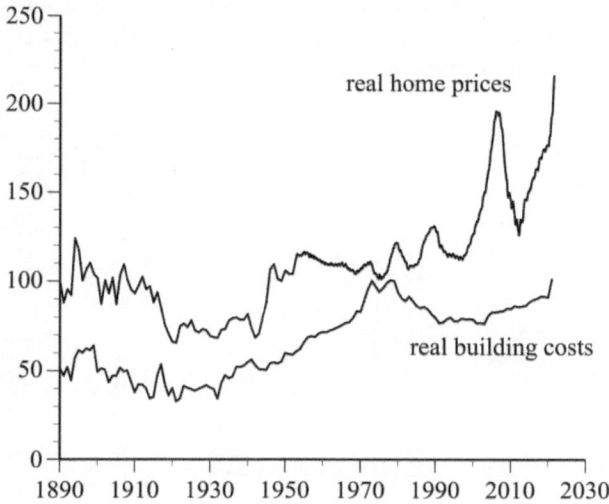

2.21 An Internet company's CEO gave her Board of Directors a graph prepared by the company's finance group that showed the company's revenue over the previous seven quarters. The Board asked her to explain why revenue had fallen so much. How would you have responded?

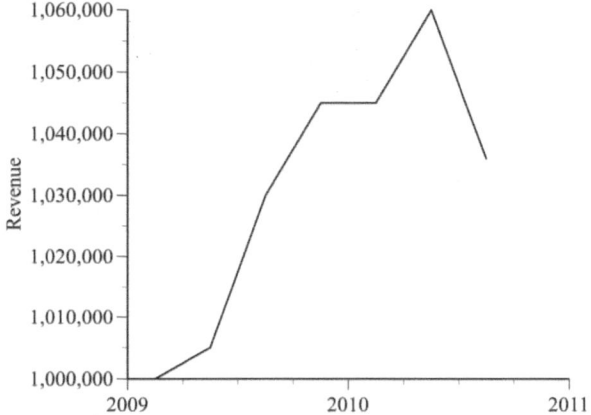

2.22 In March 2015, a professor of Psychology and Music reported that musicians in traditional musical genres (like blues, jazz, country) live much longer than do musicians in relatively new genres (like metal, rap, hip hop). The professor concluded that performing in new genres was more dangerous than going to war: "People who go into rap music or hip hop or punk, they're in a much more occupational hazard profession compared to war. We don't lose half our army in a battle." Suggest a statistical explanation for why these data are misleading.

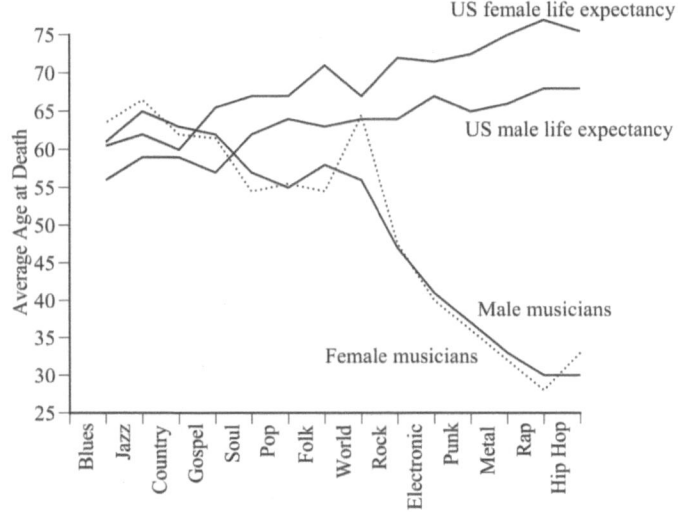

2.23 A researcher summarized a set of data with this histogram and box plot. Give two reasons why you know that the researcher made a mistake.

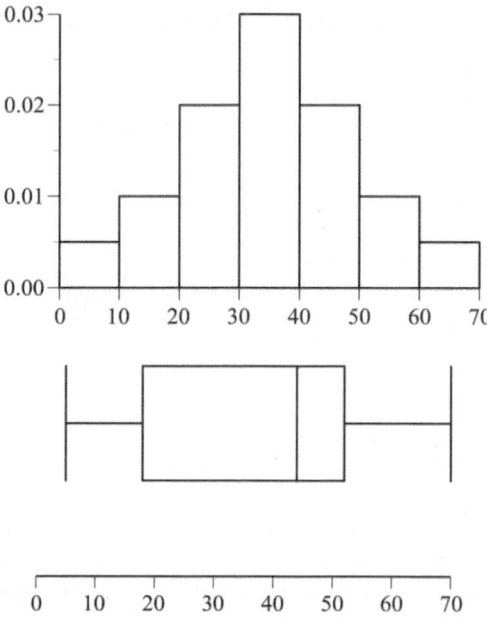

2.24 A researcher recorded annual Boston precipitation for 100 years, 1898–1997, and drew this "histogram" by calculating density as the precipitation each year divided by the total precipitation for all 100 years. Why is this not a histogram?

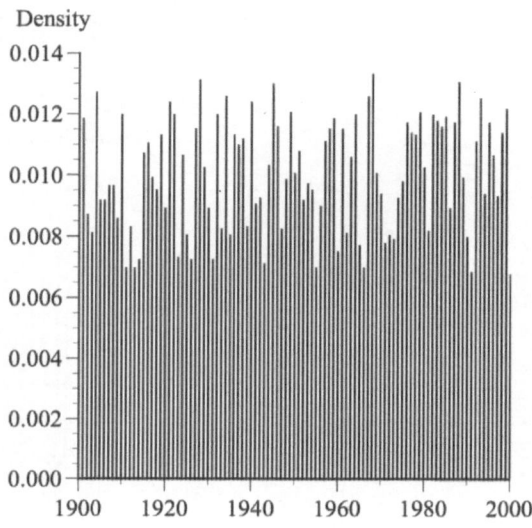

2.25 Explain how these *Wall Street Journal* graphs give the misleading impression that the improvements in math and reading scores were quite similar over this 4-year period:

Percentage Point Improvement in Test Scores

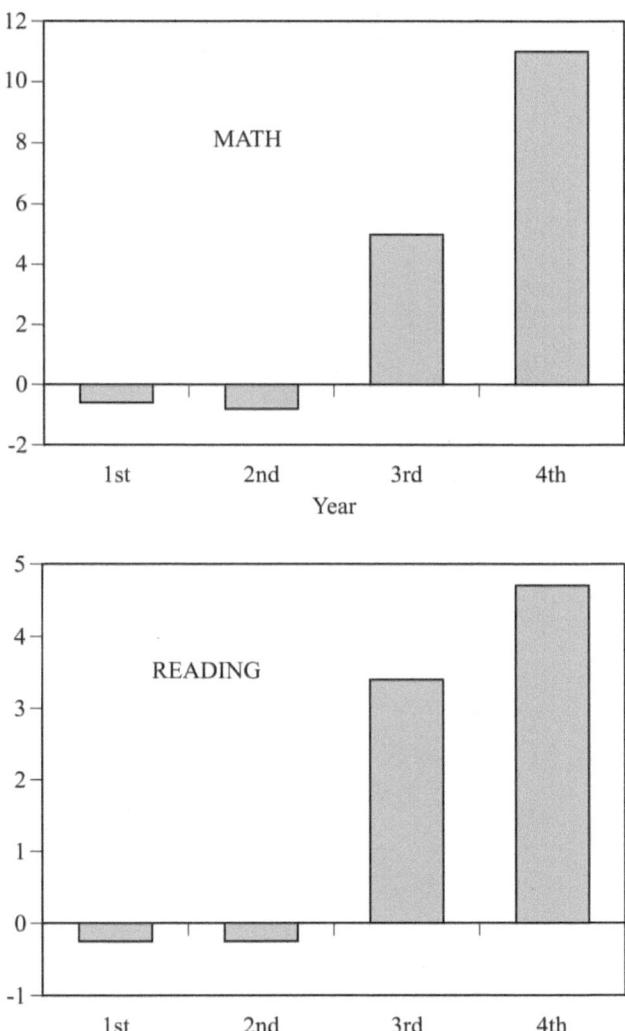

2.26 What is misleading about this graphic that accompanied a *New York Times* story about how discount fares were reducing travel-agent commissions?

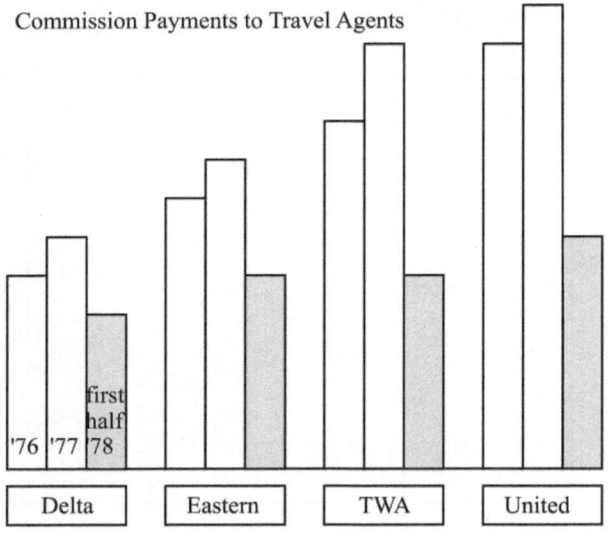

Commission Payments to Travel Agents

2.27 Identify several problems that prevent this graphic from a college newspaper from conveying useful information:

Campus Poll Reveals Little Change in Perceived Safety

How students responded when asked,
"Do you feel safe on campus?"

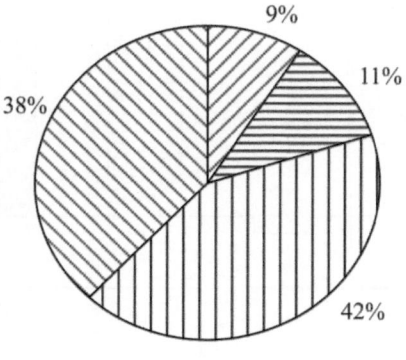

1 = not safe at all   5 = completely safe

2.28 Why is this *Washington Post* figure a misleading depiction of the fact that prices (approximately) doubled between 1958 and 1978, halving the purchasing power of a dollar?

1958 Eisenhower: $1.00

1978 Carter: 44 cents

2.29 Without doing any calculations, explain why a pie chart and a bar chart would both be poor choices for displaying these data on the number of commercial banks of different sizes.

Assets (Dollars)	Number of Banks
0 to 50 million	1,521
50 to 100 million	756
100 to 500 million	893
500 million to 1 billion	1,104
1 to 10 billion	765
10 to 50 billion	102
50 to 100 billion	16
100 to 3,500 billion	33

2.30 For each of the following studies, identify the type of graph (one histogram, side-by-side box plots, one scatter diagram, or one time series graph) that would be MOST appropriate.

a. Have test scores in an introductory statistics class risen or fallen in the past 20 years?

b. Do colleges that accept a large percentage of their students in early-decision programs have higher yields (percentage of accepted students that enroll)?

    c. Can starting salaries be predicted from college grade point averages?

    d. Are final exam scores independent of homework scores?

    e. Is there more dispersion in the starting salaries of economics majors or history majors?

2.31 For each of the following studies, identify the type of graph (one histogram, side-by-side box plots, one scatter diagram, or one time series graph) that would be MOST appropriate.

    a. Are stock returns more volatile than bond returns?

    b. Do interest rates affect home construction?

    c. Can college grade point averages be predicted from high school grade point averages?

    d. Do cigarette sales depend on cigarette prices?

    e. Is the daily precipitation in January essentially the same in Chicago, Los Angeles, and New York?

2.32 For each of the following studies, identify the type of graph (one histogram, side-by-side box plots, one scatter diagram, or one time series graph) that would be MOST appropriate.

    a. Do countries with lots of smokers have lots of lung-cancer deaths?

    b. Does the time between eruptions of Old Faithful depend on the duration of the preceding eruption?

    c. Do students who get more sleep get higher grades?

    d. Has air pollution in Dallas generally risen or fallen during the past 20 years?

    e. Could the state-by-state performance of Al Gore in the 2000 presidential election have been predicted from how well Bill Clinton did in each state in 1996?

2.33 For each of the following studies, identify the type of graph (one histogram, side-by-side box plots, one scatter diagram, or one time series graph) that would be MOST appropriate.

    a. Are final exam scores normally distributed?

    b. Does the incumbent party's chances of winning the presidential election depend on the national unemployment rate at the time of the election?

    c. Is there more dispersion in midterm or final exam scores in an introductory statistics class?

    d. Have temperatures in Los Angeles generally increased or decreased over the past 100 years?

    e. Are final exam scores higher for students who sit in the front row or in the back row?

Answers

2.1 The horizontal axis has no numbers but the equal spacing is surely wrong; *revised* should be before *original*; the numbers on the vertical axis are upside down—enrollment seems to be declining when it's actually increasing.

2.2 The figure is confusing because time is on the vertical axis and the spacing of the years is inconsistent, ranging from 2 to 6 years. Also, the data should be adjusted for increases in prices and population over this 32-year period. An improved figure shows that inflation-adjusted per capita taxes dipped during and shortly after World War II and then increased greatly to accommodate the post-World War II baby boom.

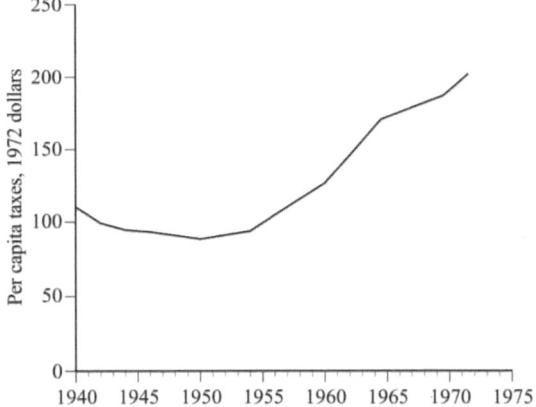

2.3 The vertical axis is upside down. When drawn in a normal fashion, the figure shows an increase in murders after the law was passed.

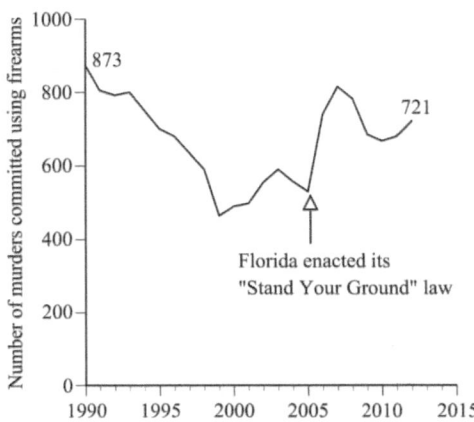

2.4 The omission of the origin from the vertical axis exaggerates the visual impression, making a 3% drop look like a 30% drop. With the origin included on the vertical axis, the figure shows that the drop in assets was minor.

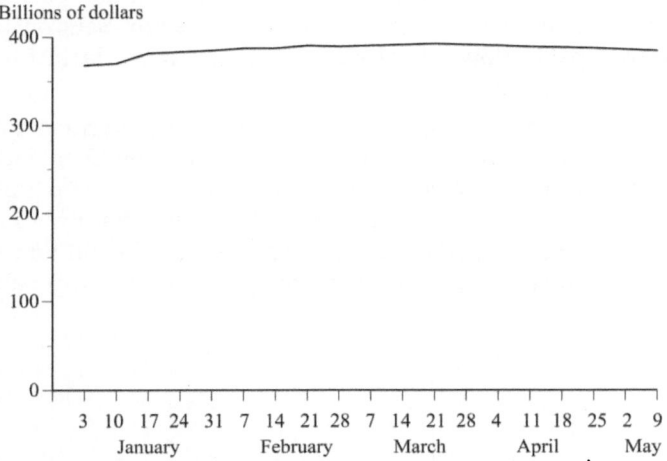

2.5 There are no units on the vertical axis so there is no way of knowing how big the savings would be. In fact, the origin was omitted in order to exaggerate the savings. It looks like taxes would be 90% lower with the Republican bill but the actual difference was only 9%.

2.6 The 1970s data only span 4 years, while the other decades have 10 years of Nobel Prizes. The NSF might have shown the number of prizes per year to account for these unequal time spans. When the full 1970s data became available, there was no sign of a decline in American Nobel Prizes.

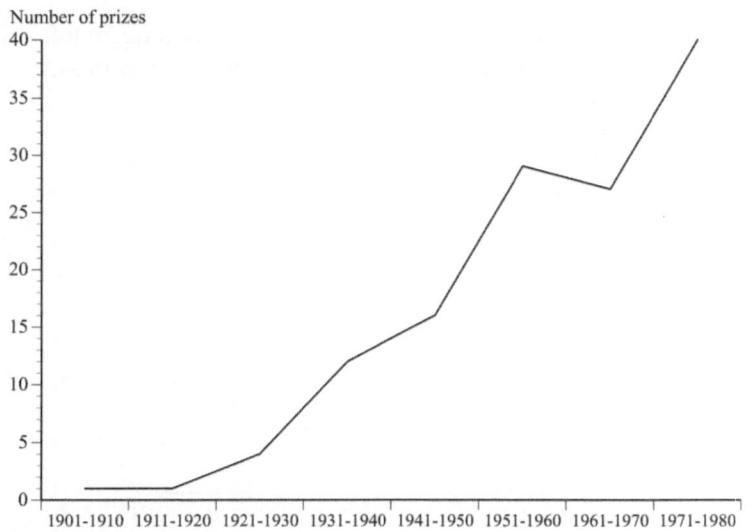

2.7 The years are not spaced correctly; the gaps between years range from 1 to 8 years. In addition, doctor income should be adjusted for the increase in the price level over this 37-year period. This figure corrects those problems and gives a very different look than the original figure.

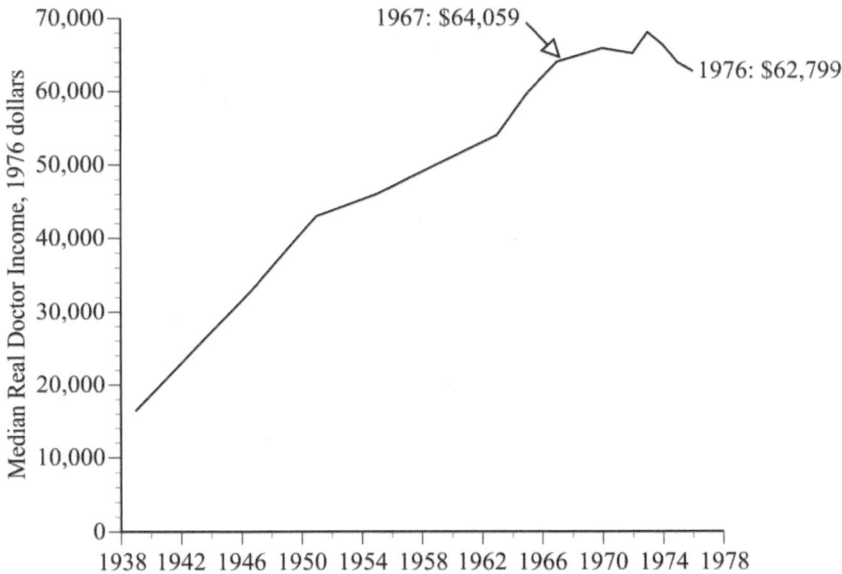

2.8 The figure compares Cornell's tuition over the 35-year period 1965–2000 with the college's ranking over the 12-year period, 1988-1999. There are two vertical axes—one for cost and one for ranking—and zero has been omitted from both axes, which exaggerates the changes in costs and ranking. Not only is zero omitted, but the entire axis and the associated numbers have disappeared, so we have no way of gauging whether the changes were large or small. Also, the *US News* ranking went down, but in college rankings, number 1 is the best! Going from, say, 17th to 13th is actually an improvement.

2.9 The omission of zero from the vertical axis magnifies the zigs and zags. Between 2:00 and 2:30, for example, the height of the line drops by 80%; yet, the actual decline in the Dow was only about .0016%.

2.10 Since the advertisement appeared in 2016, a statistician should suspect that they cherry-picked a particular 10-year period (September 6, 2001, to September 6, 2011) to support their advertising claim that gold has been a great investment. That suspicion is confirmed by considering the years before and after their cherry-picked 10-year period:

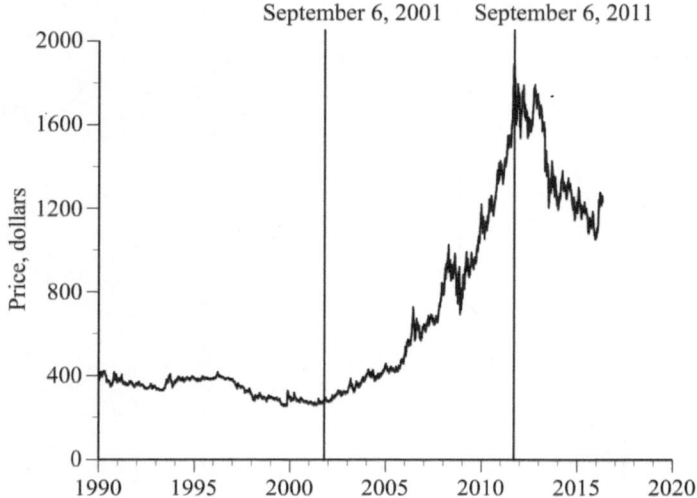

2.11 The bar widths should reflect the income-interval widths (e.g. the $15,000–$24,999 bar should be twice the width of the $10,000–$14,999 bar) and the heights should be adjusted for the different interval widths. The bars labeled "middle class" are the middle three intervals of nine arbitrary intervals, leaving 11% with lower incomes and 45% with higher incomes than the "middle class." If the middle class is instead defined as $25,000 to $74,999, it would be the middle 50%, with 25% having higher incomes and 25% lower incomes.

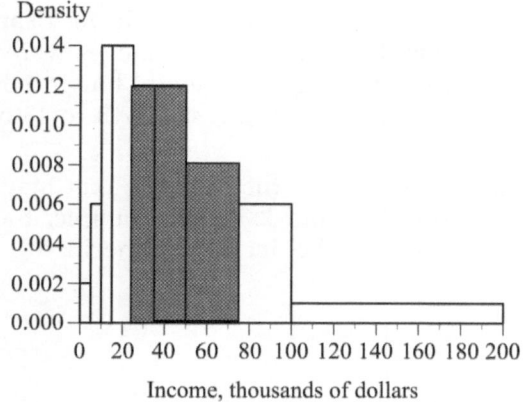

2.12 There is a 5-year difference between each of the first four bars, but a 10-year difference between the 1980 and 1990 bars. If the bars were spaced properly and a 1985 bar included, the increase over time would appear gradual, without an abrupt jump between 1980 and 1990. In addition, prices were about four times higher in 1990 than in 1965, so that $100,000 in 1990 was roughly equivalent to $25,000 in 1965. We should compare the number of 1965 families earning $25,000 with the number of 1990 families earning $100,000, taking into account the increase in the population. It is not surprising that more people have high incomes when there are more people.

   This figure fixes these problems and includes data through 2020. The 1980s are unremarkable. The periods that stand out are the 1995–2000 dot-com bubble and 2015–2020.

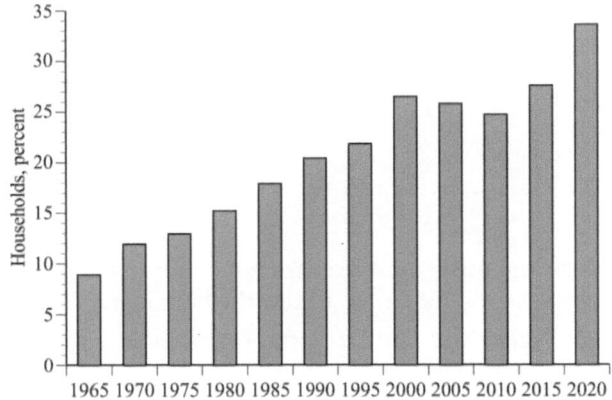

2.13 Ironically, the purpose of the figure was to illustrate that correlation is not causation. Global temperatures have been rising over time and the number of pirates has been dropping but neither is causing the other.

   The unintended lesson is that this figure is a terrific example of graphical gaffes. The vertical axis omits the origin, which exaggerates the increase in global temperatures. The horizontal axis is backwards, starting with higher numbers and then declining. The spacing of the numbers on the horizontal axis does not reflect the difference between the numbers (the distance between 17 and 400 is the same as the distance between 400 and 5000 and the distance between 5,000 and 15,000). The numbers on the horizontal axis are not in numerical order, in that 45,000 is placed between 20,000 and 35,000. The claim that there were 17 pirates in 2000 is suspiciously precise and obviously much too low. The dates on the line are evenly spaced even though the difference between the years alternates between 20 and 40. The original graph was also chart junk: using a world map as a background and seven skull-and-crossbones images for the seven data points.

2.14 The units on the horizontal axis are inconsistent, ranging from 8 months to 5 years, and there is no adjustment for inflation.

2.15 Correlation is not causation. The number of murders and Explorer's market share both fell over this 5-year period and, by using double axes with zero omitted from both, the author was able to make it seem that they were closely correlated. Here is how the appearance changes if the murder axis includes zero and the Explorer axis goes from 40% to 80%:

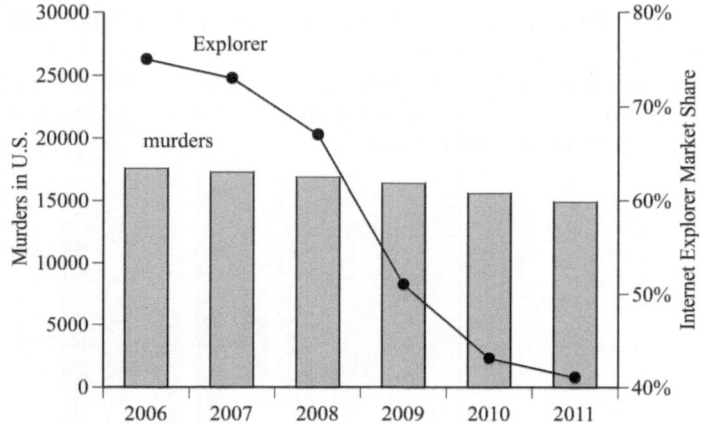

2.16 The units on the horizontal axis are not continuous and the heights are frequencies—so the total area under the bars is not one.

2.17 The use of logarithms on the horizontal axis is a problem. When the data are graphed with the number of parameters (not the log of the number of parameters) on the horizontal axis, the figure looks like this (with no tipping point):

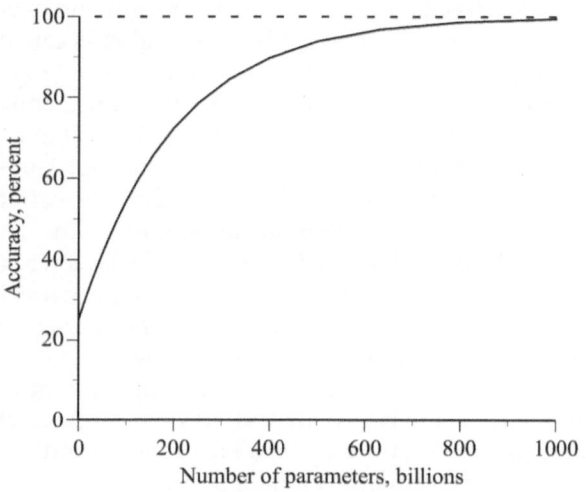

2.18 Readers are not told what the numbers on the unlabeled vertical axis mean. There is no good reason why the 12 bars should be arranged to resemble a bell curve; it would be more sensible to arrange the bars from largest to smallest. In addition, readers have to look back and forth from the graph to the legend to see which bar goes with which casino, and the similarity in the bar patterns makes it tedious and difficult to do so. The bar patterns themselves are distracting and the three-dimensional appearance is not helpful. Finally, even if we decipher the bars, it is not easy to line up the bar heights with the numerical values on the vertical axis. A simple table would be far more informative:.

Casino	Location Value
Pechanga Casino	181
Morongo Casino Resort & Spa	162
Pala Casino	153
Sobobo Casino	145
Valley View Casino	118
Casino Pauma	115
Viejas Casino & Turf Club	94
Agua Caliente	91
Cahuilla Creek Casino	69
Augustin Casino	68
Golden Acorn	50
Chumash Casino	38

2.19 Time is inexplicably on the vertical axis. This figure shows the same data, with time where it belongs—on the horizontal axis. This reversal of the axes reverses the conclusion; Yankee ticket prices did not slow down, but accelerated, after 1994. The annual rate of increase was 6% between 1967 and 1994, and 21% between 1994 and 2010.

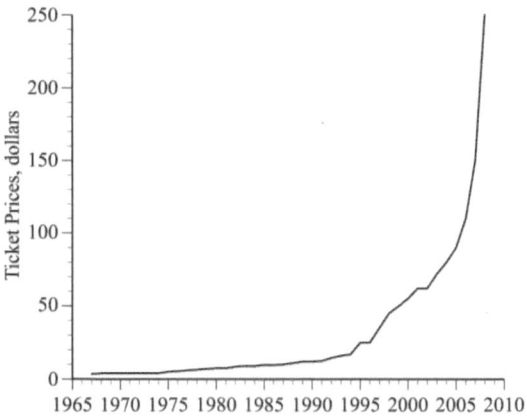

2.20 The two lines should start at the same point (e.g. 100) in order to see which increased more:

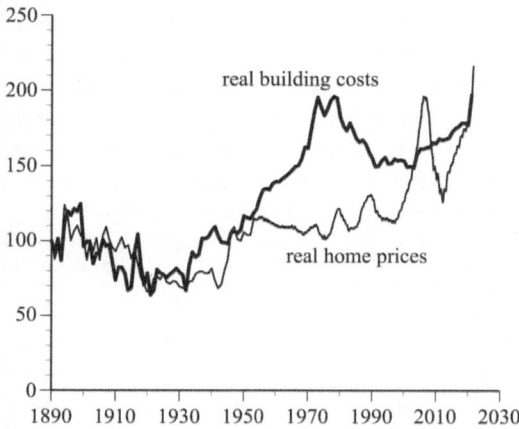

2.21 The finance group's figure omitted zero from the vertical axis and this magnified the ups and downs in the data. Revenue fell 2%, but the height of the line drops 40%.

2.22 Most people performing in new genres haven't been alive long enough to die at an elderly age. For example, hip hop began in the late 1970s. People who began doing hip hop in 1980 at age 20 and are still alive would be 57 years old in 2017. Anyone who died before 2017 would be younger than 57. People who began doing hip hop after 1980, at age 20 and died before 2017, would be even younger.

2.23 The histogram shows a symmetrical distribution, so the median should be somewhere between 30 and 40, yet the box plot shows the median to be above 40. The interquartile range in the box plot stretches from about 19 to 52, but the histogram shows that the range from 20 to 50 encompasses more than 50% of the data.

2.24 For a histogram, we should group the 100 annual observations into a small number of categories, such as 0–10 inches, 10–20 inches, and so on. The density is the fraction of the total number of years with rainfall in that interval, divided by the interval width.

2.25 The vertical axis in the math graph runs from –2 to –12, while the vertical axis in the reading graph goes from –1 to 5. By stretching the axis in the reading graph, the smaller gains are made to look comparable to those in math.

2.26 These bar graphs seem to show a large drop in commission payments to travel agents but for each airline, the first two bars are for an entire year, while the third bar is for only half a year. A doubling of these third bars would show an increase in commission payments. (Plus, the second half of the year has some important holiday travel dates, and may consequently generate even more commissions than does the first half of the year.)

2.27 The artist did not identify which answer goes with each pie slice. The caption indicates five possible responses, yet the pie has only four slices. The 11% slice looks too big in relation to the 9% slice. We are told that "1 = not safe at all" and that "5 = completely safe," but we are not told the prompts for the other three numbers. We are not told the number of people surveyed (it was a minuscule 27). We are told that this poll "reveals little change in perceived safety," but we are not shown the results of the earlier poll.

2.28 Halving the height and width of the dollar reduces its area by 75%, not half. (The original *Washington Post* figure compared purchasing power during five presidential terms in this misleading way.)

2.29 With eight different categories, a pie chart would be cluttered; in addition, it would not give a visual picture of the width of each interval, only the number of banks in each interval. A bar chart would be misleading because the intervals have different widths, so the heights of the wide-interval bars would be misleadingly large.

2.30 The most appropriate graphs are
   a. time series graph
   b. scatter diagram
   c. scatter diagram
   d. scatter diagram
   e. side-by-side box plots

2.31 The most appropriate graphs are
   a. side-by-side box plots
   b. scatter diagram
   c. scatter diagram
   d. scatter diagram
   e. side-by-side box plots

2.32 The most appropriate graphs are
   a. scatter diagram
   b. scatter diagram
   c. scatter diagram
   d. time series graph
   e. scatter diagram

2.33 The most appropriate graphs are
   a. histogram
   b. scatter diagram
   c. side-by-side box plots
   d. time series graph
   e. side-by-side box plots

3

Misleading Data

3.1 A 2001 study of four Philadelphia neighborhoods concluded that children living in neighborhoods with more books in their libraries and public schools received better grades in school. A subsequent $20 million grant from the William Penn Foundation funded a 5-year project to improve 32 neighborhood libraries in order to "level the playing field" for all children and families in Philadelphia. As a statistician, why are you unpersuaded by this study?

3.2 A popular advice columnist, Ann Landers, once asked her readers, "If you had it to do over again, would you have children?" She received 10,000 responses, of which 70% said no, they would not. Why should we interpret this poll cautiously?

3.3 A Health Maintenance Organization (HMO) survey found that more than 90% of its members are satisfied. Identify two important biases that may affect the results. Is the reported satisfaction rate biased upward or downward?

3.4 Why are you skeptical of this advertisement from the Motley Fool investment advisory service that appeared on Instagram on January 5, 2021, and boasted of its stock market recommendation for Zoom?

> *We recommended **Zoom** in the COVID Crash on March 19, 2020. Since this recommendation, Zoom is up **401.77%** from **$123.77** to **$497.27** as of 10/12/20.*

3.5 As a statistician, what is the error in this reasoning?

> *An Austrian physician is reputed to have estimated vulnerability of soldiers in World War I from hospital records, and to have concluded that injuries to limbs were most dangerous and injuries to the head least dangerous because so many hospitalized soldiers were wounded in the limbs and so few in the head.*

3.6 In World War II, the British Royal Air Force (RAF) planned to attach heavy plating to its airplanes to protect them from German fighter planes and land-based anti-aircraft guns. The protective plates weighed too much to cover an entire plane, so the RAF collected data on the location of bullet and shrapnel holes on planes that returned from bombing runs. Most of the holes were on the wings and rear of the plane, and very few on the cockpit, engines, or fuel tanks—suggesting that the protective plates should be put on the wings and rear. Do you agree?

DOI: 10.1201/9781003630159-3

3.7 Red Lion Hotels ran full-page advertisements claiming that, "For every 50 business travelers who try Red Lion, a certain number don't come back. But 49 of 50 do." The basis for this claim was a survey of people staying at Red Lion, 98% of whom said, "they would usually stay in a Red Lion Hotel when they travel." Use a numerical example to explain why the survey results do not prove the advertising claim.

3.8 American Express and the French tourist office sponsored a survey that found that most visitors to France do not consider French to be especially unfriendly. The sample consisted of "1,000 Americans who have visited France more than once for pleasure over the past two years." Why is this survey biased?

3.9 A *New York Times* article argued that one reason many students don't graduate from college within 6 years of enrolling is that they choose to "not attend the best college they could have." For example, many students with a 3.5 high school grade-point average could have gone to the University of Michigan Ann Arbor, which has an 88% graduation rate, but choose instead to go to Eastern Michigan, which has a 39% graduation rate.

   a. What flaws do you see in the conclusion that these students would have a better chance of graduating if they had gone to the University of Michigan instead of Eastern Michigan?

   b. What kind of data would we need to draw a valid conclusion?

3.10 A 1950s study found that married men were in better health than men of the same age who never married or were divorced, suggesting that the healthiest path for a man is to marry and never divorce.

   a. Explain why there might be sampling bias in this study.

   b. Suppose that marriage is generally bad for a man's health. Explain how it could still be true that men who marry and stay married are in better health than: (i) men who don't marry; and (ii) men who marry and divorce.

   c. What would have to be done for a controlled experiment?

3.11 In 2011, a group of researchers reported that Australians who watch 6 hours of television a day die, on average, 5 years younger than people who don't watch any television. Taking into account people's lifetime television habits, the researchers concluded that outlawing television would increase life expectancy by about 2 years. As a statistician, how would you challenge their study?

3.12 A researcher reported that the number of diet-related articles in the popular press totaled 60 during the entire year 1979, but totaled 50 in the single month January 1989. As a statistician, what other information would you want to see before concluding that the popular press was much more concerned about dieting in 1989 than in 1979?

3.13 Napoleon Hill's *Think and Grow Rich*, originally published in 1937, sold more than 100 million copies and is the twelfth best-selling book of all time, behind *The Da Vinci Code* and ahead of *Harry Potter and the Half Blood Prince*. *Think and Grow Rich* is actually based on Hill's previous book, *The Law of Success*, written in 1925, which reported the results of Hill's interviews with 45 millionaires in order to identify the characteristics this group shared in common. As a statistician, what problems do you see with this study?

3.14 *The Millionaire Next Door* by Thomas J. Stanley and William D. Danko, promises to reveal the "surprising secrets of America's wealthy." To uncover these secrets, the authors sent an 8-page survey, a return envelope, and a dollar bill to 3,000 Americans living in wealthy neighborhoods. A total of 1,115 surveys were returned, including 385 from people with a reported net worth of more than $1 million. The authors then identified 7 characteristics that these 385 wealthy people had in common. What is the biggest problem with this study?

3.15 Based on interviews with 700 couples who had been married for decades, Karl Pillemer, professor of human development at Cornell University, identified several secrets for a long-lasting marriage, including generosity, kindness, and effective communication. Even if this were a random sample of couples who had been married for decades, what is the main problem with these data? What would have to be done to avoid this problem?

3.16 A 2001 Harvard study concluded that students at colleges that ban all alcohol "were 30% less likely to be heavy episodic drinkers and more likely to abstain from alcohol." What problem do you see in the interpretation of this study?

3.17 As a statistician, why are you suspicious of US Senator Ted Cruz's statement during an interview on Late Night with Seth Meyers: "Many of the alarmists on global warming, they've got a problem because the science doesn't back them up. In particular, satellite data demonstrate for the last 17 years, there's been zero warming."

3.18 Many people become real estate agents, hoping to earn big commissions, but then quit after a few disappointing years. A survey of the agents at one real estate firm found that there was a positive relationship between the annual commissions and the number of years working as a realtor. The company's president argued that these results are evidence that new agents shouldn't give up. Why are these data potentially misleading?

3.19 Which statistical principle do you think Nassim Taleb used when he argued that estimates of the chances of success in the restaurant business are unreliable because, "The cemetery of failed restaurants is very silent."

3.20 Consider a hypothetical study that compares the health of a group of office workers with the health of astronauts returning from space missions. If the study showed no significant health differences, should we conclude that space missions have no health risks for astronauts?

3.21 In April 2021, after COVID-19 vaccines were approved for public use, Wisconsin Senator Ron Johnson said that he was "highly suspicious" of the government's vaccination campaign and that there was "no reason to be pushing vaccines on people." In May 2021 he said that the vaccines may not be safe: "We are over 3,000 deaths within 30 days of getting the vaccine." Why is his evidence unpersuasive?

3.22 The Public Health Service has data on the number of divorces, categorized by the month in which the marriage ceremony was performed. Thus 6.4% of divorces involve couples who were married in January. Do these data show that people who marry in June or August are more likely to get divorced than are people who marry in January?

Month Married	Fraction of Divorces	Month Married	Fraction of Divorces
January	.064	July	.087
February	.068	August	.103
March	.067	September	.090
April	.073	October	.078
May	.080	November	.079
June	.117	December	.087

3.23 A 2016 survey of adults between the ages of 20 and 100 concluded that elderly people have the greatest satisfaction with life. What statistical problem might make this conclusion misleading?

3.24 In October 2001, Reuters Health reported that

> In a study of more than 2,100 secondary school students, researchers found that boys who used computers to do homework, surf the Internet, and communicate with others were more socially and physically active than boys who did not use computers at all.
>
> On the other hand, boys who used computers to play games tended to exercise less, engage in fewer recreational activities, and have less social support than their peers.
>
> The findings, published in the October issue of the Journal of Adolescent Health, suggest that parents should monitor how their sons use their computers—and not just how much time they spend in front of the screen, lead author Dr. Samuel M.Y. Ho from the University of Hong Kong, told Reuters Health.

What problem do you see with this conclusion?

3.25 In 2015, a widely circulated news story was titled, "Selfies have killed more people than sharks this year." The subtitle was "Humans: still the world's deadliest predator." The story backed up its claim that humans are the deadliest predator with a report of 8 deaths from shark attacks in 2015 compared to 12 deaths related to taking ill-advised selfies, including people falling off cliffs, crashing their cars, being hit by trains, and shooting themselves while posing with guns. The story concluded that "not only does the likelihood of being killed by a shark pale in comparison to the deadliness of selfies, it's also a lot lower than the number of deaths caused by dog attacks and home renovations. In fact, pretty much everything you do today (particularly if it involves a car) is more likely to kill you than a shark." Why is this report misleading?

3.26 Shere Hite sent detailed questionnaires (which, according to Hite, took an average of 4.4 hours to fill out) to 100,000 women and received 4,500 replies, 98% saying that they were unhappy in their relationships with men. A *Washington Post-ABC News* poll found that 93% of the women surveyed considered their relationships with men to be good or excellent. How do you explain this difference?

3.27 A representative of the Getty Museum recently boasted that its internship program, which provides full-time summer jobs at the museum, was increasing the number of students choosing careers at museums and other nonprofit visual arts institutions: "43% of our former interns are now working at museums and other nonprofit visual arts institutions." Why does this statistic not necessarily prove its conclusion?

3.28 A state study compared the traffic fatality rates (number of fatalities per miles driven) on highways with 55, 65, and 75 miles per hour speed limits. They found that highways with a 75 miles per hour speed limit had the lowest fatality rate and highways with a 55 miles per hour speed limit had the highest fatality rate. Traffic fatalities could evidently be reduced by raising speed limits, perhaps because people pay more attention when they are driving fast. What is the biggest statistical problem with this study?

3.29 A study of 115 cats that had been brought to New York City veterinary hospitals after falling from high-rise apartment buildings found that 5% of the cats that fell from 9–32 stories died, while 10% of the cats that fell from lower heights died. The doctors speculated that this was because cats falling from higher heights are able to spread out and create a parachute effect. What might a statistician say?

3.30 A study found that, controlling for the number of cars and miles driven, people driving sports cars were more likely to get ticketed than were people driving minivans. Can you think of a reasonable explanation other than the police like to pick on people driving sports cars?

3.31 It was reported that,

> *Researchers at the University of Adelaide compared adverse events of St. John's wort and the antidepressant drug fluoxetine (Prozac). The team*

used information from doctors' reports to Australia's national agency
on drug safety. Between 2000 and 2013, there were 84 adverse reaction
reports for St. John's wort. There were 447 reports for Prozac.

As a statistician, what is the most important reason that you would advise caution in concluding that St. John's wort is safer than Prozac?

3.32 A Vanguard study of mutual funds operating in 2013 found that 62% beat the overall stock market during the previous 5 years. What statistical problem do you see here?

3.33 A month before the 2020 US Presidential election, Jimmy Failla, the host of Fox Across America, wrote,

Democratic presidential nominee Joe Biden looks to be ahead in most polls
but you'd never know it from driving through the Midwest and count-
ing the campaign signs on people's houses. No, by that metric, President
Trump is on pace to win in a lawn-slide....

I can report with full confidence that a recent family road trip took us
through the battleground states of Ohio, Pennsylvania, and Michigan and
in all three states the "Biden-Harris" signs were running a distant third
behind "Trump-Pence" and "Sweet Corn For Sale."

I was so taken by the phenomenon that I've been discussing it non-stop
with our nation's truck drivers on my radio show, "Fox Across America."
With less than 30 days until the election, the general consensus from call-
ers to my radio show is that Trump signs outnumber Biden signs by about
25 to 1 in all 50 states.

Give two distinct explanations why this sign poll turned out to be so wrong.

3.34 A professional outdoor guide, instructor, and naturalist recently wrote that the Grand Canyon in Arizona is the most dangerous National Park because 134 people have died there since 2010, more than in any other National Park. As a statistician, how would you question this statistic?

3.35 Average grades are generally higher in senior-level college chemistry courses than in introductory chemistry courses, suggesting that students who get bad grades in an introductory chemistry course shouldn't give up taking chemistry courses. Identify the most important statistical bias in these data and describe a (hypothetical) experiment that would avoid this problem.

3.36 A 2010 survey found that the average employees at the WD Company had worked for the company for 20 years, suggesting that the average person who takes a job with WD will work for the company for 20 years. Explain why these survey data might be consistent with a situation is which the average person who takes a job with WD works for the company for

a. less than 20 years.

b. more than 20 years.

3.37 A 2024 Subaru commercial said that 98% of the Subarus purchased in the past 10 years were still on the road. Does this statistic show that there is a 98% chance that a Subaru will last for 10 years?

3.38 A study of 20,072 emergency room admissions at a UK hospital found that patients who were admitted on public holidays were 48% more likely to die within 7 days than were patients admitted on other days. One interpretation of these statistics is that the doctors who work in emergency rooms on public holidays are less qualified and should be avoided. Provide another interpretation.

3.39 In 1989 New York state's highest court called marriage "a fictitious legal distinction." A *Wall Street Journal* pro-marriage editorial noted that "Adult single men are five times more likely to commit violent crimes than married men," suggesting that marriage is a major factor in reducing violent crime. How else might these data be explained?

3.40 In 2018 it was reported that people who eat lots of cheese have fewer strokes and less risk of cardiovascular disease. Why might this study be flawed?

3.41 In July 1983, Congressmen Gerry Studds (a Massachusetts Democrat) and Daniel Crane (an Illinois Republican) were censured by the House of Representatives for having had sex with teenage interns. A *Cape Cod Times* questionnaire asked readers whether Studds should "resign immediately," "serve out his present term, but not run for reelection," or "serve out his present term, and run for reelection." Out of a circulation of 51,000 people, 2,770 returned the survey:

Resign immediately	1,259	(45.5%)
Serve out his term but not seek reelection	211	(7.5%)
Serve out his term and run for reelection	1,273	(46%)
Undecided	27	(1%)

Why are these survey results almost surely misleading?

3.42 The Census Bureau regularly estimates the lifetime earnings of people by age, sex, and education. Using these estimates for 1983, a newspaper reported that

> 18-year-old men who receive a college education will be "worth" twice as much as peers who don't complete high school. On the average, that 18-year-old can expect to earn just over $1.3 million while the average high-school dropout can expect to earn about $601,000 between the ages of 18 and 64.

Mike O'Donnell was getting poor grades in high school and thinking of dropping out and going to work for his family's construction firm. But his father read this article and told Mike that dropping out would be a $700,000 mistake. Is Mike's dad right?

3.43 A 2010 study of a large California company identified white male employees who had been hired to entry-level positions 10 years earlier. This study found that men with noticeable southern accents had received smaller salary increases over this 10-year period than did men without southern accents. What is the biggest problem with this study?

3.44 Explain any possible flaws in this conclusion:

> *A drinker consumes more than twice as much beer if it comes in a pitcher than in a glass or bottle, and banning pitchers in bars could make a dent in the drunken-driving problem, a researcher said yesterday. Scott Geller, a psychology professor at Virginia Polytechnic Institute and State University in Blacksburg, VA, studied drinking in three bars near campus. Observers found that, on average, bar patrons drank 35 ounces of beer per person when it came from a pitcher, but only 15 ounces from a bottle and 12 ounces from a glass.*

3.45 A major California newspaper, *The Sun*, asked its readers the following question: "Do you think an English-only law should be enforced in California?" Of the 2,674 persons who mailed in responses, 94% said *yes*. Why might this be a very biased sample of the opinions of Californians?

3.46 A petition urging schools to offer and students to take Latin noted that, "Seventy per cent of English words are of Latin origin, hence Latin students have larger English vocabularies. In fact, Latin students score 150 points higher on the verbal section of the SAT than their non-Latin peers." How might a skeptic challenge this argument?

3.47 Dr. Timothy Roberts told a 2002 meeting of the Pediatric Academic Societies that a 1995–1996 government survey of 4,600 high-school students found that both boys and girls were much more likely to have smoked cigarettes, used alcohol and marijuana, had sex, and skipped school if they had body piercings. Does his study imply that we could reduce these activities by making body piercings illegal?

3.48 A recent study compared the death rates (number of deaths each year per 1,000 players) of people who play a variety of games, including Bingo (41.3), Beer Pong (1.6), and Mortal Kombat (.04). As a statistician, how would you respond to this conclusion?

> *The numbers reveal an alarming truth: Bingo may not be as safe as some people have assumed. "This is the first evidence we have seen that violent video games actually reduce the death rate," says PlayGear CEO Pete Elor, "It comes as a blow to the head for people who advocate less violent forms of entertainment."*
>
> *Lawyer Gerald Hill thinks the church and community need to take action: "When you look at the numbers, there's just no way you can get around it. Bingo is claiming lives right and left. And it doesn't just affect the Bingo community. Friends and families of those we lose to Bingo suffer even a greater loss." We can only hope that this study will cause people to think twice before engaging in such risky behavior.*

3.49 "At retirement, a person can choose to take a single lump sum payment or a fixed annual income until death. Those who choose the annual income live, on average, 2 ½ years longer than the general population. This shows that peace of mind is very important to one's health." Provide an alternative explanation.

3.50 In 2024 A *Wall Street Journal* article reported that people generally retire earlier than they had planned: "There is a big gap between how workers envision the timing of retirement and the reality for retirees." Two surveys were the basis for this conclusion. One survey asked people who were working when they planned to retire. The second survey asked retirees when they had retired. This table shows the percentage of responses in various age categories, with the actual retirement ages generally younger than the planned retirement ages. What fundamental statistical problem do you see with this study?

Retirement age	Expected vs. Actual Retirement Age	
	Expected	Actual
Less than 55	5	14
55–59	9	18
60–61	12	9
52–64	10	29
65	28	12
66–69	13	13
70 or older	15	6
Will never retire	9	

3.51 Use some specific hypothetical numbers to explain why these data do not necessarily justify this conclusion by the magazine *California Highways*:

> *A large metropolitan police department made a check of the clothing worn by pedestrians killed in traffic at night. About four-fifths of the victims were wearing dark clothes and one-fifth light-colored garments. This study points up the rule that pedestrians are less likely to encounter traffic mishaps at night if they wear or carry something white after dark so that drivers can see them more easily.*

3.52 "The most dangerous place to be is in bed, since more people die in bed than anywhere else." Do you agree?

Answers

3.1 There is self-selection bias. The availability of books may well have been a proxy for other socioeconomic factors. Children living in neighborhoods with oak trees might get better grades in school, but this doesn't mean that planting oak trees will raise grades. (The study did not consider whether the students actually read books—or climb oak trees.)

3.2 There is self-selection bias. Her readers were not a random sample of parents and those who responded were not a random sample of readers. Readers who are unhappy with their children may have been more likely to respond. A scientific random sample asking the same question found that more than 90% answered yes.

3.3 The satisfaction rate will be biased upward by two kinds of survivorship bias: people who left the HMO because they were dissatisfied and people who died because of poor medical treatment.

3.4 The dates are suspicious. It is unlikely that March 19, 2020, was the only day that Motley Fool recommended Zoom. Why, on January 5, 2021, are they talking about performance as of October 12, 2020, three months earlier? The chart below confirms that Zoom did not do well between October 12, 2020, and January 5, 2021.

In addition, we don't know how strongly they recommended Zoom, or the performance of the other stocks that they recommended at that time. Perhaps they recommended 25 stocks equally strongly and only one (Zoom) did well.

3.5 There is survivorship bias. Many people with head injuries were not taken to hospitals because they were dead.

3.6 There is survivorship bias in that the planes that were shot down, perhaps because of damage to their cockpits, engines, or fuel tanks, were not in the study.

3.7 The people surveyed are not a randomly selected group who were asked to stay at Red Lion and then tell whether they would come back.

Instead, the people surveyed chose Red Lion; people who tried Red Lion and never came back were not surveyed. Suppose 1,000 people try Red Lion and 49 decide to return and 951 do not. A survey is now made of 50 people—the 49 returnees plus 1 person trying Red lion for the first time. In this hypothetical example, 95% of the people who try Red Lion for the first time never go again, but 49 of 50 people staying at Red Lion returned.

3.8   There is self-selection bias in that those who visited France for pleasure and returned are less likely to consider French unfriendly than are the people who visited once and never returned.

3.9   a.   There is self-selection bias in that students may have valid reasons for choosing the colleges they attend. Perhaps almost all of the students who choose Eastern Michigan over Michigan do graduate (they are part of the 61%) and would not have graduated if they had gone to the University of Michigan Ann Arbor.

      b.   As with medical trials, we could take students accepted to both colleges and randomly choose which students go to which college. Fortunately, we can't do that.

3.10   a.   People who marry and divorce self-select.

      b.   Women may be less likely to marry men who are in poor health and more likely to divorce men whose health suffers the most during marriage.

      c.   A controlled study would compel randomly selected people to not marry, marry and stay married, or marry and divorce. Fortunately, this won't happen.

3.11   These data have self-selection bias. People who choose to watch TV all day may be less active, more depressed, or in worse health than are the people who have more interesting things to do and are healthy enough to do them.

3.12   January is not a random month. Many people start a new year resolving to lose weight, and the popular press takes advantage of that interest in diet-related articles. We should compare January to January, or whole year to whole year, and we should take into account changes in the size of the popular press between 1979 and 1989.

3.13   It is a retrospective study with survivorship bias. Any group of people, currently rich or poor, are bound to have some common characteristics. A valid study would identify the characteristics ahead of time and then compare two groups over time, one with these characteristics and one without.

3.14   This was a retrospective study with survivorship bias. Any group of people are bound to have some common traits, just by chance. A prospective study, in contrast, specifies the traits before the study begins and then monitors people with and without these traits ever time.

3.15 This study is tainted by survivorship bias because it excludes people who didn't stay married long enough to be surveyed. A valid study would look at newlyweds, identify which couples have traits thought to be important, and then track their marriages over time.

3.16 There is self-selection bias in that students who don't like drinking and drunks may be more likely to attend schools that ban alcohol. The study also found: "among drinkers, students at schools with a ban engaged in as much extreme drinking and experienced the same rate of alcohol-related problems as students at schools without an alcohol ban."

3.17 It is suspicious that he picked the last 17 years rather than a more natural 10, 20, or 100 years. This figure shows annual global land-ocean temperatures (degrees Celsius) back to 1880, as far back as these data go. The data are shown as deviations from what the average temperature was for the 30-year period, 1951 through 1980. There was a temperature spike in Senator Cruz's baseline year, 1998 (caused by a strong El Niño and unusually warm oceans). If Senator Cruz had used the more natural 1995 (20 years prior to his 2015 interview), the data would not have supported his claim.

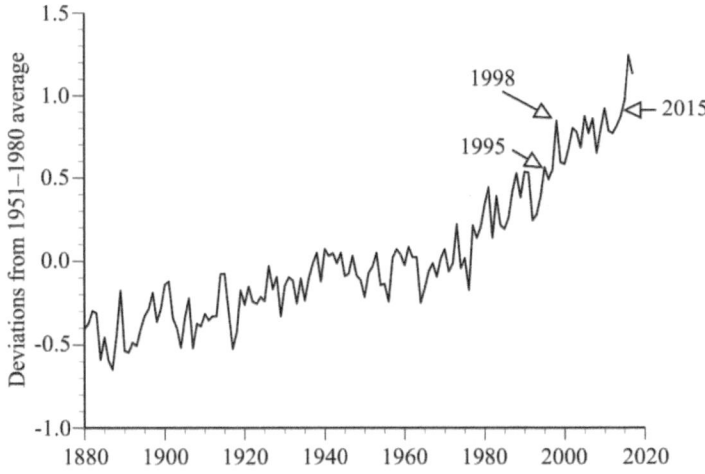

3.18 There is survivorship bias. Agents earning low commissions are more likely to quit.

3.19 There is survivorship bias. We don't know how many new restaurants have failed. (Most fail within a year; 80% within 5 years.)

3.20 The astronauts passed rigorous health tests. We would expect them, on average, to be healthier than randomly selected office workers. If they are no longer healthier after returning from space missions, that will be worrisome. There may also be survivorship bias if some astronauts did not return.

3.21 This is a *post hoc ergo propter hoc* fallacy. More than 100 million Americans had been vaccinated and the overall annual number of deaths in the United States pre-COVID is between 8 and 9 people per thousand. Applying that death rate to a month and 100 million people, we would expect about 7,000 deaths. For a more precise estimate, we should omit deaths from automobile accidents and the like, but take into account that the elderly were the first to be vaccinated. According to the CDC website,

> CDC and FDA physicians review each case report of death as soon as notified and CDC requests medical records to further assess reports. A review of available clinical information, including death certificates, autopsy, and medical records has not established a causal link to COVID-19 vaccines.

3.22 We need to know the fraction of all marriages that occurred in each month. In fact, because there are relatively more marriages in June and August (11.8% of all marriages take place in June and 11.7% in August), an examination of most marital characteristics will tend to find a preponderance of June and August marriages.

3.23 These data may well be tainted by survivorship bias in that people who became increasingly miserable as they aged died, perhaps because of poor health or suicide.

3.24 Maybe the boys who used computers to play games did so because they tended to exercise less, engage in fewer recreational activities, and have less social support than their peers, not vice versa.

3.25 To calculate the probability of being injured while taking a selfie, we need to divide 12 by the number of selfies taken (a lot), and compare this to 8 divided by the number of swims in shark-infested waters (not so many). If we apply the selfie/shark mistake to other Darwinian activities, we might conclude that selfies are more dangerous than swallowing knifes and juggling chain saws. Don't.

3.26 There is nonresponse bias. Dissatisfied people are more likely to respond to this very long questionnaire.

3.27 There is surely self-selection bias in that those who apply for these internships are likely to be interested in careers in these fields. The fact that only 43% chose such careers after an internship is worrisome.

3.28 These are observational data. Perhaps, the state chose to put 75 miles per hour speed limits on the safest highways (those that are straight and well lit, with light traffic).

3.29 There is survivorship bias in that cats that died during the fall were not brought to the hospital. Also, cat owners may have given up on cats that were still alive, but badly injured in high falls, while cat owners whose pets fell shorter distances were more optimistic.

3.30 These are not random samples. There is self-selection bias in that people who choose to drive sports cars may like to drive fast, while people who choose to drive minivans have children and drive cautiously.

3.31 Most importantly, we need to know how many people took each medication. Also, these are self-reported adverse events and we don't know how severe they were.

3.32 There is survivorship bias. They omitted funds that existed 5 years ago and are no longer around. When Vanguard included funds that existed 5 years ago and subsequently folded, the percent that beat the market fell to 46%.

3.33 There are several possible explanations: people living near truck routes may be disproportionately Trump supporters; Trump supporters may be more likely to put up signs; Trump-supporting truck drivers may be more likely to listen to this radio show and call in; and Trump-supporting truck drivers may have selective recall about the signs they saw.

3.34 We should deflate by the number of visitors.

3.35 There is self-selection bias in that students who do well in introductory chemistry courses are more likely to take senior-level chemistry courses. A controlled experiment might require all students who take introductory chemistry courses to take senior-level chemistry courses, too.

3.36 There are two offsetting biases here. We don't know which is stronger.

    a. There is survivorship bias. Suppose that most of the people who work for WD are fired after one year, but a few people stay with WD for a long time. If so, the average longevity of people still working at the company will be longer than the average longevity of people hired by the company.

    b. Current employees are not done working for WD, so the number of years they have worked there so far underestimates the total number of years they will work for the company.

3.37 Only the cars purchased 10 years ago have lasted for 10 years. Most Subarus purchased in the last 10 years were bought less than 10 years ago.

3.38 Perhaps people do not go to emergency rooms on public holidays unless they are deathly ill. Perhaps there are more serious injuries on holidays, due to more travel, excessive drinking, and reckless behavior.

3.39 There is self-selection bias. Maybe violent people are less likely to want to marry or less likely to appeal to potential spouses.

3.40 Many of those who ate lots of cheese in the past may now be deceased (survivorship bias). The cheese-eaters who survived may have lower genetic risks, exercise more, or have other traits that offset their cheese eating. It is also possible that some people who do not eat cheese today were once big cheese-eaters, but gave cheese up because of health problems. If so, they are more at risk, not because they avoid cheese, but because they once feasted on it.

3.41 The people who responded are not a random sample of readers and readers are not a random sample of voters (especially during the summer in Cape Cod). The people who have the strongest feelings about this issue are the most likely to respond. (It is striking that 27 people took the trouble to say they were undecided. The actual number of undecideds is surely much larger than this.)

3.42 There is self-selection bias. Those who go to and graduate from college are likely to be more talented and motivated than Mike—who can get a good job without going to college.

3.43 There is survivorship bias in that those workers who are still with the company are not a random sample of those hired in 2000. Some employees may have been fired, others may have taken better jobs elsewhere.

3.44 Common sense says that those who ordered pitchers were planning to drink a lot and fulfilled their objectives. Pitchers do exert psychological pressure to finish what has been paid for, but big drinkers will surely still drink a lot even if they are forced to do so using a glass or bottle.

3.45 The people who took the trouble to mail in responses are not a random sample of readers since only those who feel strongly about this issue are likely to mail in responses. The readers of this newspaper are unlikely to be a random sample of Californians—most obviously because people who are not fluent in English are unlikely to read this newspaper.

3.46 There is self-selection bias. The argument assumes that it is the taking of Latin that causes the high verbal SAT scores; it may be that the students who take Latin are above-average students with above-average language skills who would have scored 150 points above the mean even if they hadn't taken Latin.

3.47 Correlation is not causation. Perhaps the factors that lead to smoking, drinking, and so on also lead to body piercing.

3.48 There is self-selection bias. The people who play Bingo are much older than those who play Mortal Kombat.

3.49 The correlation may go in the other direction: people who are unhealthy or who have parents who died young may reasonably choose a single lump sum payment.

3.50 The survey of retirees omits people who have not yet retired and may retire later than those who have already retired. This is clearest in a comparison of the 9% of workers who said they don't plan to retire with the complete absence of people who will never retire from the survey of retirees.

3.51 Here is a contingency table with hypothetical data:

	Dark clothes	Light clothes	Total
Accident	80	20	100
No Accident	720	180	900
Total	800	200	1000

With these made-up numbers, the probability of an accident is 80/800 = .10 for those wearing dark clothes and 20/200 = .10 for those wearing light clothes—because I assumed that 80% of all pedestrians wear dark clothes. For the data to indicate that dark clothes are dangerous, it would have to be the case that fewer than 80% of all pedestrians wear dark clothes. If more than 80% wear dark clothes, then dark clothes will apparently be safer!

3.52 People who are near death are often placed in beds, but it isn't the bed that kills them.

4

Probabilities

4.1 In this carnival game, the player rolls four balls, one at a time, down a bumpy ramp, at the end of which are four boxes:

Each box is deep enough to hold all four balls and each ball is equally likely to land in any of the four boxes (this is a game of chance, not skill). The contestant pays $1 to play and wins $5 if each ball lands in a different box. What is the probability of winning the game? How much profit can the carnival expect to make, on average, per play?

4.2 In this carnival game, three balls are tossed into a tic-tac-toe arrangement of nine slots, each large enough to hold one ball:

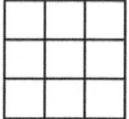

The slots are surrounded with a fence, so that each ball has to land in one of the slots. If the balls land randomly, what are the chances of a tic-tac-toe (three balls in a row horizontally, vertically, or diagonally)?

4.3 On long car trips, Mrs. Smith drives and Mr. Smith gives directions. When there is a fork in the road, his directions are right 30% of the time and wrong 70%. Having been misled many times, Mrs. Smith follows Mr. Smith's directions 30% of the time and does the opposite 70% of the time. Assuming independence, how often do they drive down the correct fork in the road? If Mrs. Smith wants to maximize the probability of choosing the correct road, how often should she follow his directions?

4.4 In a "pooled testing" of blood samples from 25 people, the 25 samples are mixed together into a batch (or pool) and tested together. If the test comes back negative, all 25 samples are known to be negative. If the test comes back positive, then the 25 people are tested individually. If the probability of a positive test result is .01, what is the expected value of the number of tests if pool testing is used?

DOI: 10.1201/9781003630159-4

4.5 Answer this letter to Ask Marilyn:

> *You're at a party with 199 other guests when robbers break in and announce that they are going to rob one of you. They put 199 blank pieces of paper in a hat, plus one marked "you lose." Each guest must draw, and the person who draws "you lose" will get robbed. The robbers offer you the option of drawing first, last, or at any time in between. When would you choose to take your turn?*

4.6 Smith loves to play squash. Andrabi has set up a challenge in which Smith will be given a prize if he can win two consecutive games in a three-game match against Andrabi and Ernst alternately, either Andrabi-Ernst-Andrabi or Ernst-Andrabi-Ernst. Assume that Andrabi is a better player than Ernst and that Smith's chances of winning a game against either player are independent of the order in which the games are played and the outcomes of other games. Which sequence should Smith choose and why?

4.7 What are the sucker's chances of winning this game?

> *Take an opaque bottle and seven olives—two green and five black. Place all seven olives in the bottle, the neck of which allows only one olive to pass through at a time. Ask the sucker to shake the bottle and then wager that he will not be able to roll out three olives without getting an unlucky green one.*

4.8 Ronald Fisher was reportedly inspired to think about statistical significance by a famous experiment involving an afternoon tea break at the agricultural research lab where he worked. He made a cup of tea for a colleague, Muriel Bristol, who refused to drink it because Fisher had poured milk into the cup before adding tea and she preferred it the other way around. Fisher doubted that she could tell the difference, so he challenged her to a test. He made eight cups of tea—four milk first and four tea first—and asked her to identify which was which after taking a sip from each cup. She got all eight correct. What is probability that she would get all eight correct if she just guessed randomly?

4.9 Galileo wrote a note on the probability of obtaining a sum of 9, 10, 11, or 12 when three dice are rolled. It had been argued that these are equally likely, because there are six ways to roll a 9 (1-4-4, 1-3-5, 1-2-6, 2-3-4, 2-2-5, or 3-3-3), six ways to roll a 10 (1-4-5, 1-3-6, 2-4-4, 2-3-5, 2-2-6, or 3-3-4), six ways to roll an 11 (1-5-5, 1-4-6, 2-4-5, 2-3-6, 3-4-4, or 3-3-5), and six ways to roll a 12 (1-5-6, 2-4-6, 2-5-5, 3-4-5, 3-3-6, or 4-4-4). Yet Galileo observed, "from long observation, gamblers consider 10 and 11 to be more likely than 9 or 12." What are the correct probabilities?

4.10 Mahjong is played with a set of 144 randomly shuffled tiles, including 16 wind tiles, consisting of 4 East tiles, 4 South tiles, 4 West tiles, and 4 North tiles. In Taiwanese Mahjong, each of 4 players is dealt 16 tiles at the start of the game. If you are one of the four players, what is the probability that

a. none of the 16 tiles you are dealt will be a wind tile?

b. none of the four players will be dealt a wind tile?

    c. you will be dealt all 16 wind tiles?

    d. the first four tiles you are dealt will be one East tile, one South tile, one West tile, and one North tile in any order?

4.11 Mahjong is played with a set of 144 randomly shuffled face-down tiles, including 8 bonus tiles (4 flowers and 4 seasons) and 136 other tiles. In Taiwanese Mahjong, each of 4 players is dealt 16 tiles at the start of the game. They then play the game by drawing tiles one at a time from the remaining face-down tiles. The game ends if there are only 16 remaining face-down tiles and no one has won yet. What is the probability that the game will end with all 8 bonus tiles among the last 16 face down tiles?

4.12 Answer this Car Talk puzzler:

> *Many years ago a prisoner was condemned to die. He was in his cell, and the warden came to visit him. He said to the prisoner that his odds of dying the next day are 100% (big damn surprise, right?). But the warden wants to increase the prisoner's chances to 50%. The warden gives the prisoner two shoeboxes, one with 50 white marbles, and one with 50 black marbles. The warden will come in blindfolded and pick a marble out of one of the boxes. White, he lives. Black, he dies.*
>
> *The marbles can be arranged in any way. All the marbles must be used in either box. Is there any way to improve his chances above 50%? If so, what is the new probability?*

4.13 You are offered the following game. A fair coin will be flipped until it lands tails. Then the game ends. If it lands tails on the first flip, you win $1. If it takes two flips, you win $2. If it takes three flips, you win $4. And so on…. What is the expected value of the payoff?

4.14 Answer this question to Ask Marilyn:

> *I recently returned from a trip to China, where the government is so concerned about population growth that it has instituted strict laws about family size. In the cities, a couple is permitted to have only one child. In the countryside, where sons traditionally have been valued, if the first child is a son, the couple may have no more children. But if the first child is a daughter, the couple may have another child. Regardless of the sex of the second child, no more are permitted. How will this policy affect the mix of males and females?*

4.15 A six-sided die with four green sides and two red sides is rolled 20 times. You choose one of these two sequences in advance and win $25 if it appears in these 20 rolls. Which sequence do you choose?

    a. RGRRR

    b. GRGRRR

4.16 In January 1991, shortly after the beginning of allied air raids against Iraq, a newspaper columnist wrote that "losses per mission were creeping toward the Vietnam and Korean level of four in every thousand. Such a level sounds good until you do the arithmetic from the pilot's point of view and realize that after 100 missions you have a one in three chance of being shot down." What is the correct probability of not surviving 100 missions, if each mission is independent with a .004 probability of being shot down?

4.17   A newspaper reported that if a defense barrier has a .15 probability of shooting down an incoming plane, then five barriers have a .75 probability of shooting down the plane. Assuming independence, what is the correct probability?

4.18   In November of 1996, it was reported that the probability of a devastating earthquake hitting California in any given year is 1/80, and that the probability of a devastating quake is consequently 1/20 in any 4-year period and 1/8 in any 10-year period. Explain why this logic is wrong and then give the correct probability of a devastating earthquake in any 10-year period if the probability of a devastating quake is 1/80 each year, independent of whether or not a devastating quake occurs in other years.

4.19   Suppose that 50-year-old women fall into two health-risk categories: 80% low-risk with a .003 probability of dying within a year and 20% high-risk with a .006 probability of dying within a year. If a company sells a randomly selected 50-year-old woman a 1-year life insurance policy for $5,000 that will pay either $1,000,000 or nothing, depending on whether the woman dies within a year,

    a.   is the expected value of the amount the insurance company will pay larger or smaller than $5,000?

    b.   for a high-risk woman, is the expected value of the payoff larger or smaller than $5,000?

    c.   for a low-risk woman, is the expected value of the payoff larger or smaller than $5,000?

    d.   what problem do you see for this insurance company?

4.20   The Scholastic Aptitude Test used to have a wrong-answer penalty. If a multiple-choice question has five possible answers, a student was given 1 point for a correct answer and a 1/4-point deduction for a wrong answer. What was the expected value of answering a question with a completely random guess? Now there is no wrong-answer penalty. What is the expected value of a random guess?

4.21   In the card game War, a standard deck of 52 playing cards is divided evenly between two players. Each player turns over a card, and the player with the higher card wins both cards and puts them at the bottom of his or her deck. (Aces are high, then kings, queens, and so on; the suit does not matter.) The game continues until one player runs out of cards. If the two cards turned over are the same rank (for example, two 7s or two jacks), it is War. Each player then plays one card face down and one card face up, and the higher face-up card wins all six cards. What is the probability that the very first play of the game will be War?

4.22 A standard 52-card deck has 13 cards in each suit: 13 spades, 13 hearts, 13 diamonds, and 13 clubs. In the card game bridge, each of the four players is dealt 13 cards. If you play one game, what is the probability that you will be dealt a perfect hand—all 13 cards of one suit?

4.23 There is a legend that the French nobleman Antoine Gombauld (the Chevalier de Mere) had won a considerable amount of money betting that he could roll at least one 6 in four throws of a single six-sided die, but was losing money on bets that he could roll at least one double-6 in 24 throws of a pair of dice. de Mere asked Pascal to analyze these games. Assuming fair dice, what is the probability of rolling

    a. at least one 6 in four throws of a single die?

    b. at least one double-6 in 24 throws of a pair of dice?

4.24 In the dice game birdcage, three standard six-sided dice are rolled and a triple occurs if the three numbers on the dice are the same; for example, 1, 1, 1.

    a. What is the probability of a triple?

    b. A $1 bet wins $3 if a triple is rolled and loses $1 otherwise. What is the expected value of this wager?

4.25 In the gambling game Chuck-A-Luck, a player can bet $1 on any number from 1 to 6. Three dice are thrown and the payoff depends on the number of times the selected number appears. For example, if you pick the number 2, your payoff is $4 if all three dice have the number 2. What is the expected value of the payoff?

Number of dice with selected number	0	1	2	3
Payoff (dollars)	0	2	3	4

4.26 Answer this question to Ask Marilyn, assuming that each birth is independent with an equal chance of being male or female: "If you have four children, they may all be of one sex, there may be three of one sex and one of the other sex, or there may be two of each. Which is most likely?"

4.27 In 1999 Sally Clark, an English solicitor, was accused of murdering her two infant sons; she said that both children had died of sudden infant death syndrome (SIDS). A pediatric professor testified for the prosecution that the probability that a child in an affluent family would suffer SIDS is 1/8500 and, therefore, the probability that two children in the family would suffer SIDS is (1/8500) (1/8500) = 1/72,250,000, or about 1 in 72 million. As an expert witness for the defense,

    a. How might you challenge the 1-in-72 million calculation?

    b. How would you challenge the prosecutor's conclusion that if the probability of two SIDS deaths is 1 in 72 million, then the probability that the Clark is innocent is 1 in 72 million?

4.28 If you roll six 6-sided dice, what is the probability you will roll the numbers 1-2-3-4-5-6, not necessarily in that order?

4.29 Long ago, the astragali (heel bones) of animals were used as dice. An astragalus of a hooved animal has four different sides. Experiments show that the probabilities of each of these four sides are .39, .37, .12, and .12. One game in ancient Greece was to roll four astragali simultaneously, with the best outcome being a "Venus," in which each of the four different sides appears. What is the probability of rolling a Venus?

4.30 Explain why this Yahoo News Story is misleading:

> *Women, much-maligned by the opposite sex for their supposed lack of ability behind the wheel, make far safer and more law-abiding drivers than their male counterparts, British officials said. Of those found guilty of all driving offenses by courts in England and Wales in 2002, 88% were male motorists.*

4.31 Explain why this is misleading:

> *In 70% of all traffic accidents, the driver is less than 10 miles from home when the accident occurs. This statistic shows that drivers have a tendency to drive incautiously when they are close to home, probably because familiar surroundings give them a false sense of security. However, the places people feel safest are the places where they are, in fact, at greatest risk of serious injury.*

4.32 Answer this question to columnist Marilyn Vos Savant:

> *My dad heard this story on the radio. At Duke University, two students had received As in chemistry in all semesters. But on the night before the final exam, they were partying in another state and didn't get back to Duke until it was over. Their excuse to the professor was that they had a flat tire, and they asked if they could take a make-up test. The professor agreed, wrote out a test, and sent the two to separate rooms to take it. The first question (on one side of the paper) was worth five points, and they answered it easily. Then they flipped the paper over and found the second question, worth 95 points: "Which tire was it?" What was the probability that both students would say the same thing? My dad and I think it's 1 in 16. Is that right?*

4.33 There are five balls in a bag—three red and two blue. If the balls are drawn out one at a time, what is the probability that the last ball is blue?

4.34 The National Society of Professional Engineers used the following sample question to promote their national junior-high-school math contest:

> *According to the Elvis Institute, 45% of Elvis sightings are made west of the Mississippi, and 63% of sightings are made after 2 p.m. What are the odds of spotting Elvis east of the Mississippi before 2 p.m.?*

The test answer was $(1 - .45)(1 - .63) = .2035$. Explain why this answer is incorrect.

4.35 In many state lotteries, the grand prize carries over to the next drawing if there is no winner. In 1992 the Virginia grand prize grew to $27 million. A ticket costs $1 and the buyer picks 6 of 44 numbers (no repeats)—winning the grand prize if these are the 6 winning numbers, not necessarily in order. What is the expected value of the payoff from buying a ticket?

4.36 The board game Settlers of Catan is played with each player taking turns rolling two standard six-sided dice and adding the numbers on the two dice. If a 7 is rolled in a four-person game, any player who holds more than seven cards loses half his or her cards. Smith holds ten cards and decides to take a chance that none of the four players will roll a 7. What is the probability that his strategy will succeed?

4.37 The board game Settlers of Catan is played with each player taking turns rolling two standard six-sided dice and adding the numbers on the two dice. To win a six-person game of Settlers on his next turn, Smith needs at least one of the six players to roll either a 5 or a 10. What are Smith's chances of winning?

4.38 Matt is a high school senior baseball player. His coach told Matt that he is among the top 5% of all high school senior baseball players and that his SAT scores place him among the top 5% of high school seniors academically. He is therefore likely to be recruited by an Ivy League baseball program because his combined athletic and academic success puts him among the top .25% of all senior high school baseball players: .05 (.05) = .0025. What's wrong with that calculation? Explain why you would either revise the .25% assessment upward or downward.

4.39 A college graduate applied to three MBA programs, and was wait-listed at all three. Because she had been told that a wait-listed student has a 1/3 chance of being accepted, she decided that she was certain to get into at least one of these three programs and did not apply to any others. (This is a true story.) If she has a 1/3 chance of being accepted by each program and their decisions are independent, what is the probability that she will get into at least one of these programs? Why might the decisions not be independent? If they are not independent, would you increase or reduce her chances of getting into at least one program?

4.40 The cost of developing the top-secret Norden bombsight in World War II was 2/3 the cost of the Manhattan Project and more than a 1/4 the production cost of all B-17 bombers used in the War. The US Army estimated that a bomb dropped from a B-17 at a 20,000-foot altitude using the Norden had a 1.2% probability of hitting within 100 feet of its target. Assuming independence, how many bombs would be needed to give a 95% probability that at least one bomb will hit within 100 feet of the target?

4.41 There are 24 students in a statistics class. The professor assigns 17 questions and uses a random number generator to select which student will answer each question. The selection is with replacement, so that a student may have to answer more than one question. What is the probability that there are no repeats, so that the 17 questions are answered by 17 different students?

4.42 A car was ticketed in Sweden for parking too long in a limited time zone after a policeman recorded the position of the two tire air-valves on one side of the car (in the 1:00 and 8:00 positions) and returned hours later to find the car in the same spot with the tire valves in the same position. The driver claimed that he had driven away and returned later to park in the same spot, and that it was a coincidence that both tire valves stopped in the same positions as before. The court accepted the driver's argument, calculating the probability that both valves would stop at their earlier positions as $(1/12)(1/12) = 1/144$ and feeling that this was not a small enough probability to preclude reasonable doubt. The court advised, however, that had the policeman noted the position of all four tire valves and found these to be unchanged, the very slight $(1/12)^4 = .0005$ probability of such a coincidence would have been accepted as proof that the car had not moved. As defense attorney for a four-valve client, how might you challenge this calculation?

4.43 Consider this doubling-up betting system. You are able to make the following wager as many times as you want: choose heads or tails on a completely fair coin flip. If you wager $X, you win $X if you are right and lose $X if you are wrong. Your doubling-up system is designed to win $1 to buy a soda, no matter what happens. Your initial bet is $1. If you win, you quit and buy your soda. If you lose, you now bet $2. If you win your second bet, you win $2, which covers the $1 loss on your first bet and allows you to quit and buy your soda. If you lose your first two bets, your third bet is for $4. If you lose that bet, your fourth bet is for $8. This doubling of the bet after every loss ensures that when you do win (and, surely, you will win eventually), you will have a net profit of $1 and can buy your soda.
Now suppose that you only have a few thousand dollars, so that if you lose 11 bets in a row, you will have lost $2,047 and cannot afford to double your 11th bet ($1,024) in order to make a 12th bet ($2,048)—you must stop betting and don't get your free soda. What is the expected value of this doubling-up system?

4.44 Two gamblers, Whale and Fish, are going to play a game of chance until one is bankrupt. Whale starts with $90,000; Fish starts with $10,000. The game they play is a fair game (e.g., calling coin flips) that each is equally likely to win. Either player can bet any amount that can be matched by the other player; for example, if Fish has $8,000, they can wager up to $8,000 on the next game. What is the probability that Whale will win?

4.45 Answer this centuries-old question that the Chevalier de Mere asked Blaise Pascal: two-evenly matched players are playing a sequence of games. The first person to win four games wins $1,000. They have played four games and A has won three games and B has won one game. At this point they are forced to stop playing. How should they divide $1,000, so that each person receives the expected value of what they would have won had they continued playing?

4.46 A king has a minter who makes 10,000 gold coins with the king's face on each coin. The minter makes 9,900 genuine gold coins and 100 false coins made of a cheaper material and puts the coins in 100 boxes. Each box has 99 gold coins and 1 false coin. The king suspects the deception and picks one coin randomly from each box to be tested.

   a. What is the probability that at least one of these 100 randomly selected coins will be a false coin?

   b. What if the king puts all 10,000 coins in a single box and randomly selects 100 coins from this box?

4.47 There are four coins—two made of gold and two made of lead. Which of these options gives you a better chance of getting a least one gold coin?

   a. There are two boxes, each containing one gold coin and one lead coin. You randomly select one coin from each box.

   b. There is one box, containing all four coins. You randomly select two coins from this box.

4.48 Bulgaria has a weekly national lottery in which players choose six different numbers from 1 through 42 and win the grand prize if their six numbers match (not necessarily in order); the six numbers are selected on live television. On September 10, 2009, the winning numbers (4, 15, 23, 24, 35, 42) were the same numbers picked the week before (though in different order). No one had the winning numbers on September 6, but a record 18 people had the winning numbers on September 10. If the game is fair,

   a. What is the probability that the six numbers picked in a lottery will be the same six numbers picked in the previous lottery (not necessarily in the same order)?

   b. If the lottery is held 1,000 times (50 times a year for 20 years), what is the probability that there will be at least one occasion where the numbers repeat?

4.49 There are 12 astrological signs corresponding to birth days, for example, Aquarius is January 20–February 18 and Pisces is February 19–March 20. Assuming birth dates are equally likely to be in each of the 12 signs, what is the probability that, in a group of four people, there will be at least two people with the same sign?

4.50 A basketball player who is fouled while attempting and missing a three-point shot will shoot three free throws that count for one point each. (If the three-point shot is made despite the foul, the player gets one free throw.) With 2 seconds left in a game, a team behind by three points passes the ball to its best shooter, who tries a three-point shot. If this player has a 40% probability of making this three-point shot and an independent 90% probability of making any one-point free throw, should the defense intentionally foul the shooter, preventing the three-point shot but giving the player three free throws?

4.51 Actuarial exams are considered the most difficult of all professional exams:

> *Actuaries are compensated very well. And just like any other well-paid profession, it takes a lot of work to get there. But unlike doctors or lawyers, actuaries need to pass a series of difficult tests called Actuarial Exams. These are very hard. Very very hard.*

Only 20% of the people who take an actuarial test pass on the first attempt. To become an actuary in the US or Canada, you need to pass seven or ten exams depending on the subfield. Using these data, Cory estimated that his chances of passing seven actuarial tests on the first try is only $.2^7 = .0000128 = 1/78,125$. What is wrong with this calculation?

4.52 The *Wall Street Journal* and *Washington Post* both reported the results of a study that estimated the probability that a 40-year-old, sober, seat-belted person driving a heavier-than-average car would have a fatal accident while making a 600-mile automobile trip. The researchers calculated this probability by multiplying the overall driver fatality rate by four risk factors. For example, the probability that a heavier-than-average car will have a fatal accident is .77 times the probability that a car of average weight will have a fatal accident. So, the overall driver fatality rate was multiplied by .77. This number was then multiplied by .68 because the probability that a 40-year-old will have a fatal accident is only .68 times the probability that a driver of average age will have a fatal accident. Similar adjustments were made for being sober and wearing a seat belt. What is wrong with this calculation?

4.53 The ancient Enneagram has nine personality types represented by the numbers 1 through 9 around a circle:

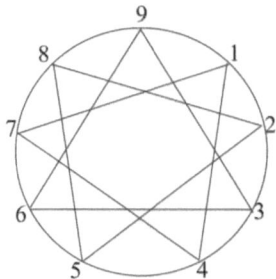

Three equilateral triangles can be drawn by connecting the points 9-3-6, 4-7-1, and 5-8-2. These three groupings correspond to three emotional states identified by modern psychological theory: attachment (9-3-6), frustration (4-7-1), and rejection (5-8-2).

If the numbers 1–9 were randomly separated into three groups of three numbers, what is the probability that one group would contain the numbers 9, 3, and 6 (not necessarily in that order), another group would contain the 4, 7, and 1, and the remaining group would contain the 5, 8, and 2?

4.54 Answer this letter to columnist Marilyn vos Savant:

> *I have a really confusing one for you. Let's say my friend puts six playing cards face-down on a table. He tells me that exactly two of them are aces. Then I get to pick up two of the cards. Which of the following choices is more likely?*
>
> a. *That I'll get one or both of the aces, or*
> b. *That I'll get no aces.*

4.55 In the traditional game Tong, two players, E and O, simultaneously reveal a hand showing one, two, or three fingers. If the sum of the fingers on the two players' hands is even, O pays $1 to E. If the sum is odd, E pays $1 to O. If each player is equally likely to show one, two, or three fingers and their choices are made independently, what is the expected value of this game for E? For O?

4.56 Alex and Blaine are having a contest shooting basketball free throws, where the first person to make a basket wins. Alex has a 2/3 chance of making a basket; Blaine has a 1/3 chance. Assuming independence, what is the probability that Blaine will win if Blaine shoots first?

4.57 Answer this Car Talk Puzzler:

RAY: Three different numbers are chosen at random, and one is written on each of three slips of paper. The slips are then placed face down on the table. The objective is to choose the slip upon which is written the largest number.

Here are the rules: You can turn over any slip of paper and look at the amount written on it. If for any reason you think this is the largest, you're done; you keep it. Otherwise you discard it and turn over a second slip. Again, if you think this is the one with the biggest number, you keep that one and the game is over. If you don't, you discard that one too.

TOM: And you're stuck with the third. I get it.

RAY: The chance of getting the highest number is one in three. Or is it? Is there a strategy by which you can improve the odds?

4.58 In 1993, 94,536,000 Butterfinger candy bar wrappers were printed, with each wrapper giving one of these prizes, depending on the message printed inside the wrapper:

Prize	Retail Value	Number of Wrappers
1993 Jeep Wrangler	$12,491.00	1
NBA team sweatshirts	34.95	100
2.1 ounce Nestle candy bar	.55	100,000
No prize	.00	94,435,899

A candy bar wrapper could be obtained by buying a Butterfinger bar or by mailing the company a self-addressed stamped envelope (one request per envelope). Is the expected value of a candy bar wrapper larger or smaller than the $.58 postage cost?

4.59 You randomly draw five cards from a well-shuffled standard deck of 52 playing cards. From these five face-down cards, you randomly select two cards. What is the probability that they are both Jacks?

4.60 A traditional game uses six peach pits that have been blackened on one side with fire. The six pits are placed in a cup, shaken thoroughly, and then inspected. The player receives five points if all six pits have a similar side face up (either all blackened or all not blackened) and gets one point if five of the six pits have a similar side face up, with one pit different. Otherwise, the player receives no points. Assuming that each pit is equally likely to land blackened or not-blackened, what is the expected value of the points for one roll?

4.61 Big 2 is played with a standard deck of 52 playing cards. The name of the game comes from the fact that the four 2s are the highest-ranked cards in the game. In a four-person game, the 52 cards are dealt equally, 13 cards to each player. In a three-person game, each player is dealt 17 cards and 1 card is set aside face down.

   a. If you are one of the players in a four-person game, what is the probability that the next hand you are dealt will have no 2s?

   b. Without doing any calculations, explain why you think that you are more likely to be dealt a hand with no 2s in a three- or four-person game.

4.62 A backgammon player needs to roll a 1 on either or both of two six-sided dice in order to leave the bar and enter the opponent's home board. On average, how many rolls of two dice will it take to get a 1?

4.63 Every box of SugarJunk cereal contains a randomly determined coupon that is equally likely to be printed with the number 1, 2, or 3. A person who collects each of the three numbers wins a prize. How many boxes are required, on average, to win the prize?

4.64 In the country OneOfEach, every woman who has babies keeps having babies until she has at least one boy and one girl, and then stops having babies. What is the average number of babies each woman has? (Assume that boy and girl babies are equally likely and do not depend on whether previous babies have been boys or girls.)

4.65 Suppose that over the rest of your life, you will call hundreds of coin flips to see who will pay for coffee, go first, or take some other action. Can you expect to win more often if you call heads every time or if you call heads half the time and tails half the time?

4.66 Determine the expected value of the wager described in this letter to "Ask Marilyn":

> *Say someone offers you the following bet: he will toss three coins all at once. If they all turn up heads, he'll give you $10. And if they all turn up tails, he'll give you $10. But if they land with either (1) two heads and a tail or (2) two tails and a head, you have to give him $5.... Should you take his bet?*

4.67 A card player claims that a thoroughly shuffled deck of 52 cards, containing four aces, usually has an ace within nine cards of the top of the deck. To test this claim, a deck is shuffled and the cards turned over one at a time. What is the probability that an ace will be among the first nine cards?

4.68 The four Simpsons pose for a family portrait. Each Simpson has a .6 probability of looking acceptable at the moment that the picture is taken. Assuming independence, what is the probability that all four Simpsons will look acceptable at the same time? How many pictures must be taken to be 80% certain that all four Simpsons will look acceptable simultaneously in at least one picture?

4.69 There are six boxes, of which five are empty and one contains a check for $10,000. Two contestants will alternate choosing boxes until one selects the box with the $10,000 check. Is there any advantage to choosing first?

4.70 Mark believes that there is a 4/5 probability that it will snow in Boston on the next New Year's Eve; Mindy believes that the probability is only 2/3. What bet could they make that would give each a positive expected value? For example, Mark pays Mindy $2 if it snows and Mindy pays Mark $5 if it doesn't.

4.71 Here is a real job-interview question: "There is a 60% chance of rain on Saturday and a 40% chance of rain on Sunday. What is the probability that it will not rain this weekend?" The interviewer said that the correct answer is $(1 - .6)(1 - .4) = .24$. The job candidate said that the correct answer is $1 - .6(.4) = .76$. What is your answer?

4.72 Answer this question to Ask Marilyn:

> *Say we're going to toss a coin repeatedly until we get one of these sequences in order: heads/tails (HT) or tails/tails (TT). If my sequence comes up first, you pay me a dollar; but if your sequence comes up first, I pay you a dollar. At that point, we stop and start all over again. Given that we're going to play this game repeatedly, which sequence would you choose?*

4.73 A, B, and C are in a three-person pistol duel, where each person takes turns with one shot. A shoots first, then B, then C, and then A again, and so on. A has a 25% chance hitting the person he aims at; B has a 75% chance of hitting the person she aims at; and C is certain to hit the person she aims at. In a two-person duel between A and B, A has a 4/13 chance of winning if she shoots first, and B has a 12/13 chance of winning if he shoots first. What is A's optimal strategy?

Answers

4.1 The first ball doesn't matter. The second ball must land in one of the three remaining boxes (which has a 3/4 probability of occurring). If it does, then the third ball must land in one of the two remaining boxes (a 2/4 probability). If it does, then the fourth ball must land in the remaining box (a 1/4 probability). Using the multiplication rule, the probability of winning is (3/4) (2/4) (1/4) = 6/64. The expected value of the payoff is (6/64) ($5) + (58/64) ($0) = $30/64. The expected value of the profit for the carnival is $1 – $30/64 = $34/64.

4.2 Think of each slot as numbered 1, 2, …, 9:

1	2	3
4	5	6
7	8	9

The number of possible combinations is 9 (8) (7)/(3) (2) (1) = 84, of which eight are winning combinations: 123, 456, 789, 147, 258, 369, 159, 357. Therefore, the probability of winning is 8/84 = .095.

Alternatively, the first ball can land in the middle, corner, or side, with these probabilities of completing a tic-tac-toe: middle + corner + side: (1/9) (8/8) (1/7) + (4/9) (6/8) (1/7) + (4/9) (4/8) (1/7) = 48/(9 × 8 × 7) = 8/84.

4.3 There are two ways they can drive down the correct road: Mr. Jones gives correct directions and his wife follows them, or he gives wrong directions and she does the opposite. The probabilities are

$$(.3) (.3) + (.7) (.7) = .58$$

In general, if P is the probability of following her husband's directions, the probability of choosing the correct road is

$$(.3)(P)+(.7)(1-P)=.7-.4P$$

which is maximized at .7 by P = 0: always doing the opposite of what he recommends.

4.4 The probability that no one in a group of 25 will test positive is equal to $.99^{25}$. Therefore, the expected value of the number of tests for a group of 25 is

$$1\left(.99^{25}\right)+(1+25)\left(1-.99^{25}\right)=1+25\left(1-.99^{25}\right)=6.55$$

4.5 It doesn't matter. Imagine each person drew a slip of paper without looking at it. Would it matter who looked first?

4.6 To win two consecutive games, Smith must win the middle game—so he should play Ernst in the second game. Letting A be the probability of defeating Andrabi and E be the probability of defeating Ernst,

Andrabi-Ernst-Andrabi: $AE+(1-A)EA = 2AE-AEA$

Ernst-Andrabi-Ernst: $EA + (1-E)AE = 2AE-EAE$

Andrabi-Ernst-Andrabi has a higher probability if E < A.

4.7 Assuming that each olive is equally likely to appear, the probability that the first olive will be black is 5/7 and, if it is black, the probability that the second olive will also be black is 4/6, since four of the six remaining olives are black. Similarly, if the first two olives are black, the probability that the third will also be black is 3/5. Thus the probability of three black olives in a row is

$$\frac{5}{7}\frac{4}{6}\frac{3}{5}=\frac{4}{14}=.286$$

4.8 Out of eight cups, Bristol has to choose four that are milk-first. Her first pick has a 4/8 probability of being correct. If it is correct, then her second pick has a 3/7 probability of being correct, and her third and fourth picks have 2/6 and 1/5 probabilities. So, the probability of getting all four correct is

$$\frac{4(3)(2)(1)}{8(7)(6)(5)}=\frac{1}{70}$$

4.9 This person's calculations mistakenly assume that each of these possible outcomes is equally likely. However, there are, for example, three ways to roll a 1, 4, and 4 (1-4-4, 4-1-4, and 4-4-1), six ways to roll a 1, 3, and 5 (1-3-5, 1-5-3, 3-1-5, 3-5-1, 5-1-3, and 5-3-1), only one way to roll 3-3-3. The correct number of ways to roll these numbers is

$$9: 3+6+6+6+3+1 = 25$$
$$10: 6+6+3+6+3+3 = 27$$
$$11: 3 + 6 + 6 + 6 + 3 + 3 = 27$$
$$12: 6 + 6 + 3 + 6 + 3 + 1 = 25$$

Galileo's theoretical calculations agree with the gamblers' empirical experience.

4.10 There are 16 wind tiles and 144 − 16 = 128 non-wind tiles.

    a. (128/144) (127/143) ... (113/129)

    b. (128/144) (127/143) ... (65/81)

    c. (16/144) (15/143) ... (1/129)

    d. (16/144) (12/143) (8/142) (4/141)

4.11 This is a multiplication problem. The probability that the first tile will not be a bonus tile is 136/144. The probability that the second tile will not be a bonus tile, given that the first tile picked was not a bonus tile is 135/143. Thus, the probability that the first two tiles are not bonus tiles is (136/144) (135/143). The probability of picking 144 − 16 = 128 tiles without any being a bonus tile is

$$\frac{136}{144} \frac{135}{143} \frac{134}{142} \cdots \frac{9}{17}$$

4.12 If the prisoner puts 1 white marble in one box and the other 99 marbles in the other box, his chances are improved to .5(1) + .5(49/99) = 148/198.

4.13 This is the St. Petersburg paradox. The probability of the game ending after one flip (tails on the first flip) is 1/2; the probability of the game ending after two flips (a HT sequence) is 1/4; the probability of the game ending after three flips (a HHT sequence) is 1/8; and so on. The expected value of the payoff is infinite but few people would pay more than a few dollars to play, demonstrating that people do not simply maximize expected return.

$$\$1\left(\frac{1}{2}\right) + \$2\left(\frac{1}{4}\right) + \$4\left(\frac{1}{8}\right) + \ldots = \infty$$

4.14 Marilyn's (correct) answer:

> *The distribution of sexes will remain roughly equal. That's because—no matter how many or how few children are born anywhere, anytime, with or without restriction—half will be boys and half will be girls. Only the act of conception (not the government) determines the sex.*

For a proof, assume that women with firstborn girls always have a second child. Suppose 100 women give birth, 50 boys and 50 girls, the 50 women with girls give birth to 25 boys and 25 girls. There are 75 boys and 75 girls.

4.15   Sequence (b) is most representative of the die because it has two greens and four reds, but it is less likely than (a) because in addition to (a), it must be preceded by a green. (From Daniel Kahneman.)

4.16   He is right. On a single mission, the probability of not being shot down is $1 - .004 = .996$. The probability of surviving 100 independent missions without being shot down is $.996^{100} = .670$. The probability of not surviving 100 missions is $1 - .670 = .330$, about 1 in 3.

4.17   The newspaper obtained the .75 probability by multiplying .15 by 5, ignoring the double counting when more than one defense barrier might be successful. By their reasoning, ten barriers would have a 10 $(.15) = 1.50$ probability of working. The correct probability is given by the subtraction rule. The probability of knocking down the plane is equal to one minus the probability that all five defense barriers fail:

$$1 - P\left[\text{all five defense barriers fail}\right] = 1 - .85^5 = .56$$

4.18   The multiplication of the number of years times 1/80 must be wrong because it implies that the probability for any period longer than 80 years is larger than 1. Assuming independence, the probability of at least one quake in a ten-year period is equal to one minus the probability of no quakes:

$$1 - (79 / 80)^{10} = .118$$

4.19   a.   The expected value of the payout is $1,000,000 (.003) + $0 (.997) = $3,000 for the women who are low-risk and $1,000,000 (.006) + $0 (.994) = $6,000 for those who are high-risk. The overall expected value is .80 ($3,000) + .20 ($6,000) = $3,600, which is less than $5,000.

      b.   For a high-risk woman, the expected value of the payoff is $6,000, which is larger than $5,000.

      c.   For a low-risk woman, the expected value of the payoff is $3,000, which is smaller than $5,000.

      d.   If only high-risk women buy policies, the insurance company will lose money because the expected value of the payoff is $6,000, which is larger than $5,000.

4.20   With a 1/4-point penalty, the expected value from guessing is $(+1)(1/5) + (-1/4)(4/5) = 0$. Without penalty, the expected value is $(+1)(1/5) + (0)(4/5) = 1/5$.

4.21 No matter what card a player turns over, the probability the other player will match it is 3/51.

4.22 The first card can be in any suit. After that the next 12 cards must be in the same suit as the first card:

$$\left(\frac{12}{51}\right)\left(\frac{11}{50}\right)\left(\frac{10}{49}\right)\cdots\left(\frac{1}{40}\right) = .00000000000629907808979643$$

4.23 Using the subtraction rule,

a. $1 - (5/6)^4 = .5177$.

b. $1 - (35/36)^{24} = .4914$.

4.24 a. $6/216 = 1/36$ because there are $6(6)(6) = 216$ possible sequences, of which six are triples. We could also reason that no matter what the number on the first die, the probability of a matching pair on the other two dice is 1/36.

b. $\mu = \$3(6/216) + (-\$1)(1 - 6/216) = -\$192/216$.

4.25 The probabilities are

Number of Dice	Probability
0	$(5/6)(5/6)(5/6) = .5787$
1	$3(1/6)(5/6)(5/6) = .3472$
2	$3(1/6)(1/6)(5/6) = .0694$
3	$(1/6)(1/6)(1/6) = .0046$

The expected value of the payoff is about an 8% loss per dollar wagered:

$$0(.5787) + 2(.3472) + 3(.0694) + 4(.0046) = .921.$$

4.26 Three of one sex is the most likely, with a probability of $8/16 = 1/2$. There are 16 possible outcomes, and there is one way to get all boys (BBBB), one way to get all girls (GGGG), four ways to get one girl and three boys (GBBB, BGBB, BBGB, and BBBG), four ways to get one boy and three girls (BGGG, GBGG, GGBG, and GGGB), and six ways to get two boys and two girls (BBGG, BGBG, BGGB, GBBG, GBGB, and GGBB).

4.27 a. The calculation assumes independence, but there may be genetic or environmental factors that contradict this assumption.

b. This (incorrectly) reverses the conditional probabilities. P [two deaths if innocent] ≠ P [innocent if two deaths].

4.28 The first die can be anything. The second die has to be one of the five remaining numbers. If it is, then the third die has to be one of the four remaining numbers. Continuing this logic, the probability is

$$1\left(\frac{5}{6}\right)\left(\frac{4}{6}\right)\left(\frac{3}{6}\right)\left(\frac{2}{6}\right)\left(\frac{1}{6}\right) = \frac{20}{1,296} = .0154$$

4.29 The probability of a specific Venus sequence is (.39) (.37) (.12) (.12). There are (4) (3) (2) (1) = 24 possible Venus sequences, since any of the four sides can appear on the first astragalus, any of the three remaining sides on the second astragalus, two on the third astragalus, and only the last side on the fourth astragalus. Thus, the probability of a Venus is 24 (.39) (.37) (.12) (.12) = .0499, almost exactly 5%.

4.30 The conclusion refers to the probability that a male (or female) driver is convicted of a driving offense, but the statistics relate to the reverse probability: the probability that a driving offense involves a male motorist. To go from one to the other, we need to know the fraction of all driving that is done by males.

4.31 The conclusion refers to the probability of having an accident if the driver is within 10 miles of home, but 70% statistic is the reverse probability: the probability of being close to home if an accident occurs. To go from one to the other, we need to know the fraction of all driving that is within 10 miles of home—surely a lot.

4.32 The first student's answer can be anything. If their excuse was untrue, the probability that the second student would select the same tire is 1/4.

4.33 Logically, if five balls are divided into two groups, one group with one ball and one group with four balls, the probability that one ball is blue doesn't depend on whether we select four of five balls and leave one or select one of five balls and leave four. The probability that the last ball is blue is 2/5 = .4.

4.34 This .2035 is the probability that, if Elvis is sighted, the sighting will be east of the Mississippi before 2 p.m. It tells us nothing about the probability of spotting Elvis, here or anywhere else. They confused P [A if B] with P [B if A].

4.35 The first number selected from 44 numbers must be one of the six chosen numbers. Given that this happens, the next number selected from 43 must be one of the five remaining chosen numbers. Continuing, the probability of winning is

$$\frac{6}{44}\frac{5}{43}\frac{4}{42}\frac{3}{41}\frac{2}{40}\frac{1}{39} = \frac{1}{7,059,052}$$

The expected value of a ticket is

$$\$27,000,000\left(\frac{1}{7,059,052}\right)+\$0\left(1-\frac{1}{7,059,052}\right)=\$3.82$$

Net of the $1 cost, the expected value is $2.82. (The prize was paid in 20 annual installment of $1,350,000 beginning immediately, so the present value is less than $27 million.)

4.36 The probability of rolling a 7 is $6/36 = 1/6$. The probability of four non-7s is $(5/6)^4 = .482$.

4.37 The probability of rolling either a 5 or 10 is $(4/36) + (3/36) = 7/36$. The probability that none of the six players will roll a 5 or 10 is $(1 - 7/36)^6 = .2733$. The probability that at least one person will roll a 5 or 10 is $1 - .2733 = .7267$. (They didn't, and Smith lost.)

4.38 This calculation assumes that athletic and academic success are independent—that the probability of being among the top 5% of baseball players is unrelated to whether one is among the top 5% academically. If athletic and academic success are positively related, the probability of being among the top 5% athletically and academically is higher than .0025; in the extreme, the probability is .05 if athletic and academic success are perfectly positively correlated. If athletic and academic success are negatively related, the probability of being among the top 5% athletically and academically is lower than .0025; in the extreme, it is .00 if athletic and academic success are perfectly negatively correlated (no one who is a top athlete is also a top student).

4.39 P [accepted by at least 1] = 1 − P [rejected by all 3] = $1 - (2/3)^3 = .7037$. They might not be independent because the three programs may look for similar traits for accepting or rejecting candidates. If so, she is more likely to be accepted by all 3 programs and more likely to be rejected by all 3. If she is more likely to be rejected by all 3, then she is less likely to be accepted by at least one because this probability is equal to 1 minus the probability of being rejected by all three.

4.40 Solving for n: $.95 = 1 - .988^n$ implies $.988^n = .05$, which implies n = ln [.05]/ln [.988] = 248.14.

4.41 The first student can be anyone. The second student must be one of the remaining 23 students, and the probability of this happening is 23/24. So the probability of selecting 17 different students is

$$P=1\frac{23}{24}\frac{22}{24}\cdots\frac{8}{24}=.000423$$

4.42 The calculation assumes that the tires rotate independently, which is debatable.

4.43 You either win \$1 or lose \$2,047. The probability of losing \$2,047 is the probability of losing 11 bets in a row: $(1/2)^{11} = 1/2,048$. The probability of winning \$1 is consequently 2,047/2,048, and the expected value of the doubling-up system is \$0:

$$\mu = (\$1)\left(\frac{2,047}{2,048}\right) + (-\$2,047)\left(\frac{1}{2,048}\right) = \$0$$

This illustrates the general principle that, if the expected value of a wager is \$0, then the expected value of any sequence of such wagers is \$0, no matter how one adjusts the size of the bets as the wagers proceed.

4.44 If it is a fair game, then each player's expected value is zero no matter what betting strategy is used. Whale's expected value is \$10,000 (P) + (−\$90,000) (1 − P), where P is Whale's probability of winning the match. The expected value is 0 if P = .9.

4.45 For B to win the \$1000, B will have to win three games in a row. The probability of this happening is $(1/2)^3 = 1/8$. Thus, A has a 7/8 chance of winning. The expected values are:

$$\text{A: } \$1000(7/8) + \$0(1/8) = \$875$$

$$\text{B: } \$1000(1/8) + \$0(7/8) = \$125$$

4.46 a. The probability that none of the 100 randomly selected coins is false is $(99/100)^{100}$. So the probability of at least one false coin is $1 - (99/100)^{100} = .644$.

   b. The probability that the first coin selected is real is 9,900/10,000. Given it is real, the probability that the second coin selected is real is 9,899/9,999. Continuing and setting the probability of at least one fake coin equal to the 1 minus the probability of all real coins:

$$1 - \frac{9,900}{10,000}\frac{9,899}{9,999}\cdots\frac{9,801}{9,901} = .6358$$

4.47 Option (b) gives you a higher probability of selecting at least one gold coin. With option (a), your selection from each box has a .5 probability of being lead. Therefore, the probability of two lead coins is .5(.5) = .25 and the probability of at least one gold coin is 1 − .25 = .75. With option (b), the probability of picking two lead coins is (2/4)(1/3) = 1/6 and the probability of at least one gold coin is 1 − 1/6 = 5/6 = .833.

4.48 a. The probability that the six numbers chosen on September 10 are the same six numbers chosen on September 6 (not necessarily in the same order) is equal to the probability that the six numbers chosen on September 10 are any pre-specified six numbers. We can use the multiplication rule:

$$\frac{6}{42}\frac{5}{41}\frac{4}{40}\frac{3}{39}\frac{2}{38}\frac{1}{37} = .000000190629$$

$$= \frac{1}{5,245,789}$$

b. We can use the subtraction rule:

$$1 - \left(1 - \frac{1}{5,245,786}\right)^{1000-1} = .00019$$

4.49 This is like the birthday paradox. First determine the probability of no matches. The first person can have any sign. The second must have one of the remaining 11 signs, the third person one of the remaining 10 signs, and the fourth person one of the remaining 9 signs. The probability of at least one match is equal to one minus the probability of no matches:

$$1 - 1\frac{11}{12}\frac{10}{12}\frac{9}{12} = .427$$

4.50 There is a .40 probability of making a three-point shot versus a $.9^3 = .729$ probability of making three free throws. Don't foul a three-point attempt.

4.51 It assumes independence.

4.52 A multiplication of the separate risk factors assumes that the risk factors are independent; for example, that being a 40-year-old is independent of being a sober driver. In fact, a 40-year-old may be more likely to be sober; a 40-year-old sober person may be more likely to wear a seat belt; and a 40-year-old, sober person wearing a seat belt may be more likely to be driving a heavier-than-average car.

4.53 The first number picked for a group can be anything. Given the first number, the second number picked for that group has to be one of the two remaining numbers that belong in that group. For example, if the first number picked for a group is 7, then the second number selected for that group has to be either a 1 or a 4. If the second number selected is a 1 or 4, then the third number selected for that group must be the third number that belongs in that group. Thus, the probability that the first three numbers picked will be in the same group is 1(2/8) (1/7). For the second group, given that the first three numbers are correctly placed in the first group, the first number picked can be any of the six remaining numbers, the second number picked has to be one of the two numbers that goes with the first number and the third number picked has to be the third number that belongs in that group. This probability is 1(2/5) (1/4). If the first six numbers are okay, then the last three numbers must be correct for the third group. Thus the overall probability is

$$1\frac{2}{8}\frac{1}{7}1\frac{2}{5}\frac{1}{4} = \frac{1}{280}$$

4.54 Marilyn arrived at the correct answer (a) by listing all the possibilities, but we can calculate the probabilities directly. The probability of no aces is equal to the probability that the first card is not an ace multiplied by the probability that the second card is not an ace, given that the first card is not an ace: (4/6) (3/5) = .40. The probability of one or two aces is one minus the probability of no aces:

$$1 - .40 = .60.$$

4.55 We can reason that the probability tree has nine equally likely branches, 1-1, 1-2, 1-3, 2-1, 2-2, 2-3, 3-1, 3-2, and 3-3, with five of the branches giving even sums. Therefore, the probability of an even sum is 5/9. E's expected value is (+$1) (5/9) + (−$1) (4/9) = $1/9. Since this is a zero-sum game, O's expected value is −$1/9.

4.56 Letting S stand for success and F for failure on any shot, Blaine's winning sequences are

Sequence	Probability
S	$(1/3) = 1/3$
FFS	$(2/3)\,(1/3)\,(1/3) = (1/3)\,(b)$
FFFFS	$(2/3)\,(1/3)\,(2/3)\,(1/3)\,(1/3) = (1/3)\,(b^2)$
FFFFFFS	$(2/3)\,(1/3)\,(2/3)\,(1/3)\,(2/3)\,(1/3)\,(1/3)$ $= (1/3)\,(b^3)$
...	

where b = (2/3)(1/3) = 2/9. The total probability is

$$P = \frac{1}{3}(1 + b + b^2 \ldots) = \frac{1}{3}\left(\frac{1}{1-b}\right) = \frac{1}{3}\left(\frac{1}{1-2/9}\right) = \frac{3}{7}$$

4.57 The probability goes up to 1/2 if you use this strategy: look at the first piece of paper and then discard it. Draw a second slip; if it is larger than the first slip, choose it; otherwise, choose the third slip.

Suppose that the highest number is on Slip 1, the second highest on Slip 2, and the lowest number is on Slip 3. Here are the possibilities.

There is a 1/3 probability that you choose Slip 1 first, discard it, and lose.

There is a 1/3 probability that you choose Slip 2 first. Half of these times, you will choose Slip 1 next, keep it, and win. The other half of the time, you chose Slip 3 second, discard it, and win.

There is a 1/3 probability that you choose Slip 3. Half of these times, you choose Slip 1 next, keep it, and win; the other half of the time, you choose Slip 2 next, keep it, and lose.

The probability of winning is $1/3 + 1/3 \, (1/2) = 1/2$.

4.58 The expected value is less than one-tenth of a penny:

$$E = \$12,491 \left(\frac{1}{94,536,000} \right) + \$34.95 \left(\frac{100}{94,536,000} \right)$$
$$+ \$.55 \left(\frac{100,000}{94,536,000} \right) + \$0 \left(\frac{94,435,899}{94,536,000} \right)$$

4.59 This is the same as drawing two cards from a deck of 52 playing cards. Using the multiplication rule, the answer is $(4/52) \, (3/51)$.

4.60 The probability of six blackened or six not blackened is $.5^6 + .5^6 = 2/64$. The probability of five blackened or five not blackened is $6 \, (.5^6) + 6 \, (.5^6) = 12/64$. Thus the expected value is $5 \, (2/64) + 1 \, (12/64) + 0 \, (40/64) = 22/64$.

4.61 a. $(48/52)(47/51)\dots(36/40)$.

b. More likely in a four-person game because there are fewer opportunities to be dealt a 2.

4.62 The probability of rolling a 1 on either or both dice is equal to 1 minus the probability of two non-1s: $1 - (5/6) \, (5/6) = 11/36$. The expected wait is 1 divided by this probability: $1/(11/36) = 36/11$.

4.63 The first number can be anything. The probability that a new box will have a different number is 2/3, so the expected wait until getting a different number is $1/(2/3) = 3/2$. After 2 different numbers, the probability that a new box will give the remaining number is 1/3, with an expected wait of $1/(1/3) = 3$ boxes. The total expected wait is $1 + 3/2 + 3 = 5.5$ boxes.

4.64 The first baby can be either a boy or girl. The expected wait until a baby of the opposite sex is $1/.5 = 2$. So, the average number of babies is 3.

4.65 It doesn't matter. Whatever strategy you use, your chances of winning a fair coin toss is .5.

4.66 The probability of three heads is 1/8, as is the probability of three tails. The probability of 2 of one and 1 of the other is 6/8. The expected value is $10(2/8) – $5(6/8) = –$1.25.

4.67 $1 – P$ [no aces] $= 1 – (48/52) (47/51)...(40/44) = .544$.

4.68 Using the multiplication rule, $.6^4 = .1296$. Now, $.8 = P$ [at least once] $= 1 – P$ [never] $= 1 – (1 – .1296)^n$. Solving, $n = \ln (.2)/\ln (1 – .1296) = 11.60$.

4.69 Each contestant is equally likely to win. Suppose each contestant chooses boxes without opening them until all boxes are chosen. Each contestant will have three boxes and an equal probability of having the box with the $10,000 check.

4.70 Because Mark has a higher probability of snow, he should bet on snow and Mindy should bet against snow. Let's make the bet that Mindy pays Mark $X if it snows and Mark pays Mindy $Y if it doesn't. From Mark's viewpoint, his expected value is $(X)(4/5) + (–Y)(1/5) > 0$ if $X/Y > 1/4$. From Mindy's viewpoint, her expected value is $(–X)(2/3) + (Y)(1/3) > 0$ if $X/Y < 1/2$. Thus, any bet with $1/4 < X/Y < 1/2$ gives both positive expected values; for example, Mindy pays Mark $1 if it snows and Mark pays Mindy $3 if it doesn't.

4.71 The probability of no rain on Saturday and no rain on Sunday is given by the multiplication rule: P [noSat and noSun] $= P$ [noSat] P [noSun if noSat]. If we unrealistically assume that Sunday rainfall is independent of Saturday rainfall, then the interviewer is correct:

$$P\left[\text{noSat and noSun}\right] = (1 - .6)(1 - .4) = .24$$

The candidate's answer is evidently

$$P\left[\text{at least 1 dry}\right] = 1 - P\left[\text{rain both days}\right] = 1 - .4(.6) = .76$$

In addition to assuming independence, the job candidate is answering the wrong question.

The correct answer is that we don't have enough information to answer this question, since we don't know the conditional probability that it will not rain on Sunday if it doesn't rain on Saturday. Some storms come and last more than a day; other times, there is no rain for several days.

4.72 Marilyn's correct answer:

I'd choose heads/tails (HT) because I'd win three out of four times! That's because there are four different sequence combinations: HH, HT, TH, and TT. If tails/tails (TT) were to appear at the very start, you'd win, but that would happen only one-fourth of the time. For TT to appear any time afterward, it would have to be preceded by H, which means that I'd win before you ever saw your sequence come up at all!

4.73 If A shoots at B and hits her, C will hit A with certainty. If A shoots at C and hits her, the probability of B defeating A is 12/13. If A deliberately misses, B's best plan is to aim at C because, given a chance, C will hit B with 100% certainty. If B hits C, then A has a 4/13 chance of defeating B. If B misses C, C will hit B and A has a 1/4 chance of defeating C. So, A's probability of winning if he deliberately misses his first shot is .75(4/13) + .25(1/4), which is greater than his probability of winning if he hits either B or C.

5

Bayes' Rule

5.1 A college senior was asked this question in an interview for a banking job. A can contains 20 coins, 19 normal coins and 1 coin that has heads on both sides. A coin is randomly selected from the can and flipped five times. It lands heads all five times. What is the probability that it is the two-headed coin?

5.2 The landmark 1964 US Surgeon General report on cigarette smoking cited these data: 6% of all deaths are due to lung cancer; 85% of the people who die of lung cancer are smokers; and 1/3 of the adult population are smokers. Compare the probabilities of dying of lung cancer for smokers and nonsmokers.

5.3 Suppose that 10% of student essays are written by ChatGPT and that a ChatGPT detection program is 90% accurate: it correctly flags 90% of ChatGPT-written essays as having been written by ChatGPT and it correctly flags 90% of human-written essays as having been written by humans. What percent of the essays flagged as written by ChatGPT were in fact written by humans?

5.4 Answer this question that doctors and residents at a Boston hospital were asked: one in a thousand people have a certain disease. A test for the disease always gives a positive result if a person has the disease and has a 5% chance of giving a false positive result for a patient who does not have the disease. What percentage of the people with positive test results actually have the disease?

5.5 In a mock courtroom experiment, 144 volunteer jurors were told that a liquor store had been robbed by a man wearing a ski mask. The police arrested a suspect near the store whose height, weight, and clothing matched the clerk's description. The ski mask and money were found in a nearby trash can. After hearing this evidence, the jurors were asked to write down their estimate of the probability that the arrested man "really did it." The average probability was .25.

Then a forensic expert testified that samples of the suspect's hair matched a hair found inside the ski mask and that only 2% of the population has hair that matches the hair found in the ski mask. After hearing this evidence, the jurors' average revised probability of the defendant's guilt was .63. What is the Bayesian posterior probability?

5.6 In 2016, two Chinese researchers reported that a computer analysis of the faces of male Chinese criminals and non-criminals correctly

DOI: 10.1201/9781003630159-5

identified 89.5% of the criminals as criminals and correctly identified 93% of the non-criminals as non-criminals. If .36% of the Chinese male population is criminal, what is the probability that a person labeled criminal by the computer is actually a criminal?

5.7   Some cars turn out to be "lemons," with many defects. Unfortunately, the quality of a new car cannot be determined until the car has been driven several hundred miles. The unlucky purchaser of a lemon may try to unload it, by resale to someone else. Suppose that 10% of all new cars are lemons and that 90% of all lemons and 5% of all non-lemons are offered for sale within the first year of ownership. If so, what is the probability that a randomly selected car from among those offered for sale within the first year is a lemon?

5.8   A letter to Ask Marilyn began: "As professors of statistics, we found your response to the drug-testing question perplexing and, indeed, incorrect." The question was,

> *Suppose we assume that 5% of people are drug-users. A test is 95% accurate, which we'll say means that if a person is a user, the result is positive 95% of the time; and if she or he isn't, it's negative 95% of the time. A randomly selected person tests positive. Is the individual highly likely to be a drug user?*

In fact, Marilyn calculated the probability correctly. What is the correct probability?

5.9   To see who serves first in their Thursday squash games, Player A spins the racket while Player B guesses whether the racket logo will stop face up or face down. Player B initially believes that there is only a 5% chance that A cheats when he spins the racket. But after A wins the first five times he spins the racket, B isn't so sure. Assuming that A will always win if he cheats and has a 50% chance of winning if he doesn't cheat, what is B's revised probability that A cheats?

5.10 In three careful studies, polygraph experts tested several persons, some known to be truthful and the others known to be lying, to see if the experts could tell which were which. Overall, 83% of the liars were pronounced "deceptive" and 57% of the truthful people were judged "honest." Using these data and assuming that 80% of the people tested are truthful and 20% are lying, what is the probability that a person pronounced "deceptive" is in fact truthful? What is the probability that a person judged "honest" is lying?

5.11 A magician claims she has practiced coin flips millions of times and can flip heads 75% of the time. She shows you a coin which appears to be a normal coin before she begins her demonstration. You think there are three possibilities:

a.   She will flip a normal coin and has a 75% chance of flipping heads.

b. She will flip a normal coin and has only a 50% chance of flipping heads, but she is hoping for lucky flips.

c. She switched the coin you saw for a coin with heads on both sides.

Before she begins her demonstration, your personal probabilities are P [a] = .2, P [b] = .3; and P [c] = .5. After she flips three heads in a row, what is your revised probability that she is flipping a normal coin and does have a 75% chance of flipping heads?

5.12 Approximately 1.5% of Americans are schizophrenic. Computed axial tomography (CAT) scans show brain atrophy in 30% of people who have been diagnosed as schizophrenic and in only 2% of people diagnosed as not schizophrenic. In the 1982 trial of John Hinckley for the attempted assassination of President Ronald Reagan, the defense attorney tried to present evidence that a CAT scan of Hinckley had shown brain atrophy, thereby indicating that Hinckley was schizophrenic. Is a CAT scan showing brain atrophy persuasive evidence that a person is schizophrenic?

5.13 Epic claimed that sepsis predictions made by its AI Epic Sepsis Model (ESM) are 76%–83% accurate, but there were no independent tests until a medical team examined the hospital records of 38,455 patients at Michigan Medicine, of whom 2,552 (6.6%) experienced sepsis. ESM generated a (correct) sepsis alert for 843 of these 2,552 patients and (correctly) did not generate a sepsis alert for 29,775 of the 35,903 patients who did not have sepsis. What fraction of ESM's sepsis alerts was false positive?

5.14 For a woman who gives birth at age 35, the probability of having a baby suffering from Down syndrome is 1/270. A test of the amniotic fluid in the mother's uterus is virtually 100% accurate in predicting Down syndrome, but is expensive and can cause a miscarriage. A study of the effectiveness of an inexpensive blood test that does not risk miscarriage found that in 89% of the Down-syndrome cases, the test gave a positive reading, while in 75% of the cases without Down syndrome the test gave a negative reading. Of those cases where there is a positive reading, what fraction is false positive?

5.15 Tay-Sachs disease is a genetic disorder that occurs in roughly 1 in 320,000 US babies. Suppose that a genetic test is 99% accurate (a positive result) for babies that have Tay-Sachs and 98% accurate (a negative result) for babies that do not have Tay-Sachs. If a test comes back positive, what is the probability the baby has Tay-Sachs?

5.16 A new theory groups people into one of three personality types (X, Y, or Z) and assumes that one-third of the population is in each group. A large empirical study asked people this question, "Would you rather go to the beach or the mountains for a two-week vacation?," and found that 70% of the Type X people answered "Beach," as did 50% of the Type Y people, and 30% of the Type Z people.

If Jill Jones answers "beach," what is the probability that she is a Type X personality?

5.17 "Five out of ten people who have this disease die from it. It is lucky you came to me; I've had five patients and they all died." Suppose that there are three kinds of doctors: excellent (10%), good (80%), and incompetent (10%). With an excellent doctor, every patient lives; with a good doctor, half live and half die; and with an incompetent doctor, every patient dies. Given this doctor's track record (five of five died), what is the probability that he/she is incompetent?

5.18 Suppose that 20% of the people who are given breathalyzer tests for intoxication are legally intoxicated and 80% are not, and that the test gives a positive reading 85% of the time when a person is intoxicated, and gives a negative reading 80% of the time when a person is not intoxicated. If the test comes back positive, what is the probability that the person who was tested is, in fact, not intoxicated?

5.19 Suppose that 10% of all investment advisers are experts and 90% are guessers. Each expert has a .6 probability of making a correct prediction whether the stock market will beat the bond market over the course of a year; a guesser has a .5 probability of being correct. Each adviser's chances of being correct are independent of the other advisers and also independent of whether they were correct in earlier years. If an adviser is correct 5 years in a row, what is the probability that this adviser is an expert?

5.20 A 1960s study found that in 3,379 criminal jury trials where the jury reached a verdict, the judge agreed with the jury's verdict in 2,696 cases (79.8%)—with both judge and jury convicting in 2,217 cases (65.6%) and both acquitting in 479 cases (14.2%). In 604 cases (17.9%), the jury acquitted but the judge would have convicted; in 79 cases (2.3%), the jury convicted but the judge would have acquitted.

a. Who was more likely to convict, the judge or the jury?

b. In those cases where the jury acquitted, what was the probability that the judge would convict?

c. In those cases where the judge acquitted, what was the probability that the jury would convict?

5.21 Israel has a high COVID-19 vaccination rate, yet on August 15, 2021, 58% of the Israelis hospitalized for COVID-19 were fully vaccinated—suggesting that vaccinations are ineffective or even harmful. Why is this 58% number misleading from a purely statistical standpoint? What statistical technique would you use to solve this problem?

5.22 Here's a stylized example of how polls can be used to predict election outcomes. Suppose that there are only two possibilities: Cameron is preferred by either 51% or 49% of all voters, and our prior probability is that each situation is equally likely. A poll of 1,000 voters finds that 510 (51%) of those surveyed prefer Cameron. What is the posterior probability that 51% of all voters prefer Cameron?

5.23 Suppose that 90% of the people who enjoyed the television show *Homicide: Life on the Street* also enjoyed the show *The Wire*, while 20% of the people who did not enjoy *Homicide* did enjoy *The Wire*. If 40% of all people enjoyed *Homicide*, what is the probability that a randomly selected person who enjoyed *The Wire* also enjoyed *Homicide*?

5.24 Fifteen percent of US households live below the poverty line. In a third of all US households, a woman is the sole income provider. In 60% of poor households, a woman is the sole income provider. Of those households in which a woman is the sole income provider, what fraction is poor?

5.25 A farmer sends out 10,000 cartons of eggs every day. Two inspectors independently glance at each carton, looking for cracked eggs, and each inspector has a .70 probability of noticing when a carton has cracked eggs. They never reject a carton that has no cracked eggs. What is the probability that a carton with cracked eggs will be noticed by at least one inspector? If 30% of the inspected cartons have cracked eggs, what is the probability that a carton that is passed by both inspector has cracked eggs?

5.26 It has been argued that diagonal earlobe creases for adult males indicate the presence of coronary heart disease. It has been estimated that 5% of all adult males have coronary heart disease and that 47% of those with coronary heart disease have diagonal earlobe creases, while 30% of those with no heart disease have earlobe creases. What is the probability that an adult male with an earlobe crease has heart disease?

5.27 In the US criminal defendants are presumed innocent until they are proven guilty beyond a reasonable doubt because it is thought "better to let nine guilty people go free than to send one innocent person to prison." Assume that 90% of all defendants are guilty, 90% of the guilty defendants are convicted, and 90% of the innocent defendants are set free. What percent of the people convicted are innocent? Of those people set free, what percent are guilty?

5.28 In Craig v. Boren (1976), the US Supreme Court considered whether important government objectives were served by the sex distinction in an Oklahoma statute that prohibited the sale of 3.2% beer to males under the age of 21 and to females under the age of 18. Among the evidence considered in this case were the following data on persons arrested in Oklahoma for driving under the influence (DUI) during the last four months of 1973: 92% of those arrested were male; 8% of the males arrested were under the age of 21; and 5% of the females arrested were under the age of 21. Assume a hypothetical population of 10,000 DUI arrests and construct a contingency table with the columns showing sex (male or female) and the rows showing age (under 21 or older). Do these data indicate that age and sex are strongly related or largely unrelated?

5.29 The Oakland Athletics, winners of 103 games during the regular season, were heavily favored to defeat the Cincinnati Reds in the 1990 World Series. After the Reds won the series in four straight games, several Athletics said that the series had been a fluke and that they were still convinced that they were the best team in baseball. One speculated that if they played 100,000 games against the Reds, the Athletics would win 60,000.

Assume that baseball games are Bernoulli trials with the Athletics having a probability p of winning a game against the Reds and that there are two possible values for p: .60 and .50. Before the 1990 World Series, the prior probabilities are $P[p = .60] = .70$ and $P[p = .50] = .30$. Use Bayes' theorem to determine the revised probability that $p = .60$ after the Athletics lose four of four games to the Reds.

5.30 A routine examination discovers a lump in a female patient's breast. Only 1 of 100 such lumps turns out to be malignant, but, to be safe, the doctor orders a mammogram X-ray test. If the lump is malignant, there is a .8 probability that the mammogram reading will be positive; if the lump is benign, there is a .9 probability that the reading will be negative. The test comes back positive and the doctor orders a second test, which comes back negative. Assuming the test results to be independent, what is your estimate of the probability that this lump is malignant?

Answers

5.1 Using Bayes' rule with HH representing the two-headed coin, HT a normal coin, and 5heads being 5 heads in a row:

$$P[\text{HH if 5heads}] = \frac{P[\text{HH}]P[\text{5heads if HH}]}{P[\text{HH}]P[\text{5heads if HH}] + P[\text{HT}]P[\text{5heads if HT}]}$$

$$= \frac{.05(1)}{.05(1) + .95)(1/32)} = .627.$$

5.2 Using Bayes' rule,

$$P[C \text{ if } S] = \frac{P[C \text{ and } S]}{P[S]} = \frac{P[C]P[S \text{ if } C]}{P[S]} = \frac{.06(.85)}{1/3} = .153$$

$$P[C \text{ if } noS] = \frac{P[C \text{ and } noS]}{P[S]} = \frac{P[C]P[noS \text{ if } C]}{P[noS]} = \frac{.06(.15)}{2/3} = .0135.$$

Using a contingency table with a total population of 3,000:

	Smoker	Nonsmoker	Total
Lung cancer	153	27	180
No lung cancer	847	1,973	2,820
Total	1000	2,000	3,000

$$P\left[\text{lung cancer if smoker}\right] = 153 / 1,000 = .153.$$

$$P\left[\text{lung cancer if nonsmoker}\right] = 27 / 2,000 = .0135.$$

Not all smokers die of lung cancer, but they are ten times more likely to die of lung cancer than are nonsmokers.

5.3 This is a Bayes' rule problem:

$$P[\text{Human if } +] = \frac{P[\text{Human}]P[+ \text{ if Human}]}{P[\text{Human}]P[+ \text{ if Human}] + P[\text{GPT}]P[+ \text{ if GPT}]}$$

$$= \frac{.9(.1)}{.9(.1) + .1(.9)} = .5.$$

5.4 The most common answer was 95%. Bayes' rule gives the correct answer of about 2%:

$$P[D \text{ if } +] = \frac{P[D]P[+ \text{ if } D]}{P[D]P[+ \text{ if } D] + P[noD]P[+ \text{ if } noD]}$$

$$= \frac{.001(1)}{.001(1) + .999(.05)} = .0196.$$

5.5 Using G for guilty and HM for hair match, and a .25 prior probability of guilt, Bayes' rule implies that the posterior probability, taking the hair match into account, is .943:

$$P[G \text{ if } +] = \frac{P[G]P[]HM \text{ if } G}{P[G]P[HM \text{ if } G] + P[notG]P[HM \text{ if } notG]}$$

$$= \frac{.25(1)}{.25(1) + .75(.02)} = .943.$$

5.6   Using Bayes' rule with this notation: + is labeled criminal, − is labeled non-criminal, C is actually criminal, and NC is actually non-criminal

$$P[G \text{ if } +] = \frac{P[C]P[+\text{if } C]}{P[C]P[+\text{if } C] + P[NC]P[+\text{if } NC]}$$

$$= \frac{.0036(.895)}{.0036(.895) + .9964(.07)} = .044.$$

(The .36% figure is probably wrong since the authors call it "the crime rate," which is probably the fraction of the population convicted of crimes every year, not the fraction of the male population that has ever committed a crime.)

5.7   Let L signify lemon, NL not lemon, and S offered for sale:

$$P[L \text{ if } S] = \frac{P[L]P[S \text{ if } L]}{P[L]P[S \text{ if } L] + [NL]P[S \text{ if } NL]}$$

$$= \frac{.1(.9)}{.1(.9) + .9(.05)} = .667.$$

5.8   Using Bayes' rule, the probability that a person who tests positive is actually a drug user is only .50:

$$P[D \text{ if } +] = \frac{P[D]P[+\text{if } D]}{P[D]P[+\text{if } D] + P[noD]P[+\text{if } noD]}$$

$$= \frac{.05(.95)}{.05P(.95) + .95(.05)} = .50.$$

5.9   Using Bayes' rule with C = cheating and W5 = 5 wins,

$$P[C \text{ if } W5] = \frac{P[C]P[W5 \text{ if } L]}{P[C]P[W5 \text{ if } C] + P[NotC]P[W5 \text{ if } notC]}$$

$$= \frac{.05(1)}{.05(1) + .95(1/32)} = .627.$$

5.10 Using a contingency table with 1,000 people tested and quotation marks identifying the expert's assessment,

	"Honest"	"Deceptive"	Total
Truthful	456	344	800
Lying	34	166	200
Total	490	510	1,000

The probability that a person pronounced "deceptive" is in fact truthful is 344/510 = .6745. The probability that a person judged "honest" is lying is 34/490 = .0694.

A federal law enacted in 1988 prohibits the use of polygraph tests to screen job applicants or to test an employee unless there is other evidence of a specific misdeed by this employee. Exemptions are provided for businesses that involve public health and safety; for example, pharmaceutical companies and security-guard firms. Polygraph results are not considered admissible evidence in federal courts and in about half of the US states.

5.11 Using Bayes' rule, your revised probability is

$$P[a \text{ if } 3H] = \frac{P[a]P[3H \text{ if } a]}{P[a]P[3H \text{ if } a] + P[b]P[3H \text{ if } b] + P[C]P[3H \text{ if } c]}$$

$$= \frac{.2(.75^3)}{.2(.75^3) + .3(.5^3) + .5(1.0)} = .1357.$$

Your probability drops from .2 to .1357, which is reminiscent of Thomas Paine's argument that, "Is it more probable that nature should go out of her course, or that a man should tell a lie?"

5.12 Using Bayes' rule, it is far from certain that a person with brain atrophy is schizophrenic:

$$P[S \text{ if } A] = \frac{P[S]P[A \text{ if } S]}{P[S]P[A \text{ if } S] + P[notS]P[A \text{ if } notS]}$$

$$= \frac{.015(.3)}{.015(.3) + .985(.02)} = .186.$$

Alternatively, the table below shows the calculations for the hypothetical case of 10,000 randomly selected Americans who are given CAT scans

	Brain Atrophy	No Brain Atrophy	Total
Schizophrenic	45	105	150
Not schizophrenic	197	9,653	9,850
Total	242	9,758	10,000

We expect 150 to be schizophrenic, with 45 (30%) showing brain atrophy, and 9,850 to be not schizophrenic, with 197 (2%) showing brain atrophy. Thus, of the 242 people who show brain atrophy, 45 (18.6%) are schizophrenic.

5.13 Using Bayes' rule, with "S" meaning sepsis, "+" meaning that ESM generated a sepsis alert and "−" meaning it did not:

$$P[\text{noS if} +] = \frac{P[\text{noS}]P[+\text{if noS}]}{P[\text{noS}]P[+\text{if noS}] + P[\text{S}]P[+\text{if S}]}$$

$$= \frac{.934\left(1 - \dfrac{29,775}{35,903}\right)}{.934\left(1 - \dfrac{29,775}{35,903}\right) + .066\left(\dfrac{843}{2,552}\right)} = .88.$$

A two-way table can also be used:

	Epic+	Epic−	Total
Sepsis	843	1,709	2,552
No sepsis	6,128	29,775	35,903
Total	6,971	31,484	38,455

The false positive frequency is 6,128/6,971 = .88.

5.14 This question can be answered either with Bayes' rule or a contingency table. Here is a contingency table with a total of 27,000 women who give birth at age 35, of whom (1/270) 27,000 = 100 will suffer from Down syndrome:

	Positive Reading	Negative Reading	Total
Down	89	11	100
No Down	6,725	20,175	26,900
Total	6,814	20,186	27,000

Of the 100 Down-syndrome babies, the blood test will give a positive reading in .89 (100) = 89 cases and a negative reading in the remaining 11 cases. Of the 26,900 babies not suffering from Down syndrome, the blood test will give a negative reading in .75 (26,900) = 20,175 cases and a positive reading in the remaining 6,725 cases.

$$P[X \text{ if } B] = \frac{P[B]P[B \text{ if } X]}{P[X]P[B \text{ if } X] + P[Y]P[B \text{ if } Y] + P[Z]P[B \text{ if } Z]}$$

$$= \frac{(1/3).7}{(1/3).7 + (1/3).5 + (1/3).3} = .467$$

Of the 6,814 cases with positive readings, a stunning 6,725/6,814 = .987 are false positives.

(The false negative rate is 11/20,186 = .0005; the main benefit of this blood test is that it can screen out many women from an amniotic-fluid test that risks a miscarriage. For those who get a positive blood-test result, a follow up amniotic-fluid test can be done.)

5.15 Using Bayes' rule,

$$P[\text{TS if } +] = \frac{P[\text{TS}]P[+\text{if TS}]}{P[\text{TS}]P[+\text{if TS}] + P[\text{noTS}]P[+\text{if noTS}]}$$

$$= \frac{\left(1 - \frac{1}{320,000}\right).99}{\left(1 - \frac{1}{320,000}\right).99 + \left(\frac{319,999}{320,000}\right).02} = .000154.$$

5.16 This is a Bayesian problem:

$$P[\text{X if B}] = \frac{P[\text{B}]P[\text{B if X}]}{P[\text{X}]P[\text{B if X}] + P[\text{Y}]P[\text{B if Y}] + P[\text{Z}]P[\text{B if Z}]}$$

$$= \frac{(1/3).7}{(1/3).7 + (1/3).5 + (1/3).3} = .467.$$

Not asked: P [Y if beach] = .333 and P [Z if beach] = .20.

5.17 Using a contingency table,

	All Five Died	**Some Survived**	**Total**
Incompetent	10	0	10
Good	(1/32)80	(31/32)80	80
Excellent	0	10	10
Total	10 + (1/32)80	10 + (31/32)80	100

Of the 10 + (1/32)80 cases where all five patients died, 10 died because the doctor was incompetent. Therefore the probability that this is an incompetent doctor is 10/(10 + (1/32)80) = .80.

5.18 Using Bayes' rule with S = sober and D = intoxicated,

$$P[\text{S if } +] = \frac{P[\text{S}]P[+\text{if S}]}{P[\text{S}]P[+\text{if S}] + P[\text{D}]P[+\text{if D}]}$$

$$= \frac{(.8).2}{(.8).2 + (.2).85} = .4848.$$

5.19 Using a contingency table with 1,000 advisers,

	Five Correct	Not Five Correct	Total
Expert	$100(.6^5)$	$100 - 100^*.6^5$	100
Guesser	$900(.5^5)$	$900 - 900(.5^5)$	900
Total	$100(.6^5)$	1,000	1,000
	$+ 900(.5^5)$	$-(100(.6^5) + 900(.5^5))$	1,000

The probability that a person who gets 5 out of 5 correct is an expert is $100(.6^5)/(100(.6^5) + 900(.5^5)) = .2166$.

5.20 The easiest way to answer these questions is to set up a contingency table:

	Jury Convict	Jury Acquit	Total
Judge convict	2217	604	2821
Judge acquit	79	479	558
Total	2296	1083	3379

a. The judge was more likely to convict, $2,821/3,379 = .835$ versus $2,296/3,379 = .679$.

b. P [judge convict if jury acquit] $= 604/1083 = .558$.

c. P [jury convict if judge acquit] $= 79/558 = .142$.

5.21 We are interested in the reverse probabilities: a comparison of the probability of being hospitalized for the vaccinated and the unvaccinated. We would use Bayes' rule to determine the relevant probabilities.

5.22 Using Bayes' rule (with 51%A shorthand for 51% of all voters and 51%S shorthand for 51% of those surveyed):

$$P[51\%A \text{ if } 55\%S] = \frac{P[51\%A]P[51\%S \text{ if } 51\%A]}{P[51\%A]P[51\%S \text{ if } 51\%A] + P[49\%A]P[51\%S \text{ if } 49\%A]}$$

$$= \frac{.5\binom{1000}{510}.51^{510}.49^{490}}{.5\binom{1000}{510}.51\%^{510}.49^{490} + .5\binom{1000}{510}.49^{510}.51^{490}}$$

$$= \frac{.5(.025230)}{.5(.025230) + .5(.011335)}$$

$$= .69.$$

5.23 We can use Bayes' rule, letting H mean enjoyed *Homicide* and W mean enjoyed *The Wire*:

$$P[H \text{ if } W] = \frac{P[H]P[W \text{ if } H]}{P[H]P[W \text{ if } H] + P[notH]P[W \text{ if } notH]}$$

$$= \frac{.4(.9)}{.4(.9) + .6(.2)} = .75.$$

5.24 This probability can be calculated from Bayes' rule or from a contingency table. To simplify calculations with the one-third woman sole income provider (WSI), assume a total population of 300. Of these, 100 (one-third) are WSI and 45 (15%) are poor. Of the 45 households that are poor, 27 (60%) are WSI. The remaining numbers in the table can be filled in from these data.

	WSI	Not WSI	Total
Poor	27	18	45
Not poor	73	182	255
Total	100	200	300

Thus P[poor if WSI] = 27/100 = .27.

5.25 The probability that a carton with cracked eggs will be noticed by at least one inspector is 1 minus the probability that neither notices: $1 - .3^2 = .91$. The probability that a passed carton has cracked eggs is given by Bayes' rule:

$$P[C \text{ if } +] = \frac{P[C]P[+ \text{ if } C]}{P[C]P[+ \text{ if } C] + P[notC]P[+ \text{ if } notC]}$$

$$= \frac{.3(.3^2)}{.3(.3^2) + .7(1)} = .37.$$

5.26 Using Bayes' rule with H for heart disease and EC for earlobe crease, the presence of a diagonal earlobe crease increases the probability of heart disease from 5% to 7.6%:

$$P[H \text{ if } EC] = \frac{P[H]P[EC \text{ if } H]}{P[H]P[EC \text{ if } H] + P[noH]P[EC \text{ if } noH]}$$

$$= \frac{.05(.47)}{.05(.47) + .95(.30)} = .076.$$

5.27 We can use a contingency table with 100 people brought to trial, of whom 90 (90%) are guilty. Letting 90% of the 10 innocent defendants be set free and 90% of the 90 guilty defendants be convicted, the complete table is

	Convicted	Set Free	Total
Innocent	1	9	10
Guilty	81	9	90
Total	82	18	100

Thus, the answers to the questions asked are:

Of those people convicted, $1/82 = .012$ are innocent.

Of those people set free, $9/18 = .50$ are guilty.

5.28 Here is a contingency table.

	Male	Female	Total
Under 21	736	40	776
21 or Older	8,464	760	9224
Total	9,200	800	10,000

P [male] = .92; P [male if under 21] = $736/776$ = .948 and P [male if 21 or older] = $8,464/9224$ = .918. If age and sex are independent, these should all be equal. They are (approximately) all equal. The Supreme Court threw out Oklahoma's sex distinction because, although males were arrested far more than females, this was true by roughly constant amounts in each age interval. Thus, there was no empirical basis for distinguishing between males under the age of 18 and males under the age of 21.

5.29 First, we calculate the probabilities of Oakland losing 4 of 4 if $p = .7$ and if $p = .5$: $P[x = 0$ if $p = .6] = .4^4 = .0256$ and $P [x = 0$ if $p = .5] = .5^4 = .0625$. Now we can use Bayes' rule:

$$P(p = .6 \text{ if } L4) = \frac{P(p = .6)P(L4 \text{ if } p = .6)}{P(p = .6)P(L4 \text{ if } p = .6) + P(p = .5)P(L4 \text{ if } p = .5)}$$

$$= \frac{.7(.0256)}{.7(.0256) + .3(.0625)} = .489.$$

If, before the World Series, you thought there was a .70 probability that the Athletics were the better team, then you should revise this probability down to .489 after the World Series.

5.30 This is a Bayesian problem using binomial probabilities. The order in which the tests are taken doesn't matter because the test results are independent. Letting M be malignant, B be benign, and +/− be one positive and one negative result, Bayes' rule is

$$P[M \text{ if } +/-] = \frac{P[M]P[+/-\text{if } M]}{P[M]P[+/-\text{if } M] + P[B]P[+/-\text{if } B]}$$

$$= \frac{.01(.32)}{.01(.32) + .99(.18)} = .0176.$$

where

$$P[+/-\text{if } M] = \binom{2}{1}.8^1.2^1 = .32$$

$$P[+/-\text{if } B] = \binom{2}{1}.1^1.9^1 = .18.$$

$$P[1 \text{ of } 2 \text{ positive if } M] = \binom{2}{1}.8^1.2^1 = .32$$

$$P[1 \text{ of } 2 \text{ positive if } B] = \binom{2}{1}.1^1.9^1 = .18.$$

This problem can also be answered in two steps. The first test, with the positive mammogram result, implies

$$P[M \text{ if } +] = \frac{P[M]P[+\text{if } M]}{P[M]P[+\text{if } M] + P[B]P[+\text{if } B]}$$

$$= \frac{.01(.8)}{.01(.8) + .99(.1)} = .075.$$

For the second test, the prior probability is now .075:

$$P[M \text{ if } -] = \frac{P[M]P[-\text{if } M]}{P[M]P[-\text{if } M] + P[B]P[-\text{if } B]}$$

$$= \frac{.075(.2)}{.075(.2) + .925(.9)} = .176.$$

Bayes' rule is internally consistent, in that the answer is the same whether we consider the two tests together or analyze them one after the other.

6

Monty Hall Problems

6.1 On the classic television show *Let's Make a Deal*, a contestant is asked to choose one of three doors—with a grand prize behind one door and goats behind the other two doors. After the contestant picks a door, the host, Monty Hall, does what he always does by showing a goat behind a door that was not chosen and asking if the contestant wants to switch doors.

What do you advise?

6.2 What is wrong with the answer to this Car Talk Puzzler?

> *Monty Hall. Let's Make a Deal. Doors 1, 2, and 3... You select doors without knowing what is behind them. One of the doors has a good prize behind it, the others have crummy prizes. Monty says, "You picked door #1, but you don't know what's behind it. I've shown you what's behind door #2, and it is a bunch of stale Eskimo pies... Would you like to switch to door #3?" What do you do, and why?*

> *Answer: The answer is, you should switch. Monty Hall is always going to show you a crummy door. Right? So you have a 50/50 chance of getting a good door, if you switch. So statistically, if you do this enough times, by switching doors, you up the chance of getting a good outcome.*

6.3 There are four doors: one door with a $10,000 prize hidden behind it and three doors with bags of goldfish. You choose Door 3. No matter which door you choose, the host will then show you a bag of goldfish behind a door you did not choose (say, Door 1) and ask if you wish to switch your choice to one of the two remaining doors. What is the probability of winning the $10,000 prize if you do switch?

6.4 There are three boxes: a box containing two gold coins, a box containing two silver coins, and a box containing one gold coin and one silver coin. You choose a box randomly and choose a coin randomly from this box. If it is a gold coin, what is the probability that it came from the box with two gold coins?

6.5 A box has three drawers. One drawer contains two gold coins; another drawer contains two lead coins; and the third drawer contains one gold coin and one lead coin. You pick a drawer at random and a coin from that drawer at random. It is a gold coin, which is set aside and you do not get to keep. Now, your options are to either (a) take the other coin in

DOI: 10.1201/9781003630159-6

the drawer that you chose; or (b) randomly pick a coin from one of the two drawers you did not choose. If you choose Option (a), what is the probability that you will get a gold coin?

6.6   Five coins are placed in a leather bag. One coin is heads on both sides; one coin is tails on both sides, and three coins are heads on one side and tails on the other side. A coin is drawn randomly from the bag and, without anyone looking at it, is flipped in the air. It lands with a heads side up. What is the probability that the other side of the coin is tails?

6.7   Three-card Monte: three cards are placed in a bag. One card is red on both sides, one is green on both sides, and one is red on one side and green on the other side. You draw a randomly selected card out of the bag and see that it is green on the front side. The operator of the game says, "We know that this is not the double-red card. We also know that it could be either green or red on the other side. I will bet $5 against your $4 that it is, in fact, green on the other side." Can the operator profit from such bets without cheating?

6.8   Two cards are placed in a leather bag. One card is red on both sides; the other card is red on one side and blue on the other side. If you pick a card randomly and look at a side randomly and see that it is red, what is the probability that the other side of the card is also red?

6.9   A wooden box has four cards inside. One card is red on both sides; one card is blue on both sides; and two cards are red on one side and green on the other side. If you pick a card randomly and look at a side randomly and see that it is red, what is the probability that the other side is green?

6.10 Three prisoners, A, B, and C, are in separate cells and sentenced to death. The governor has randomly selected one of them to be pardoned. The warden knows which one, but is not allowed to tell. Prisoner A begs the warden to let him know the identity of one of the others who will be executed. "If B is to be pardoned, give me C's name. If C is to be pardoned, give me B's name. And if I'm to be pardoned, flip a coin to decide whether to name B or C." The warden tells A that B is to be executed. Prisoner A is pleased because he believes that his probability of surviving has gone up from 1/3 to 1/2, as it is now between him and C. Prisoner A tells C the news, who is also pleased because he reasons that A still has a 1/3 chance of being pardoned, so his chance has gone up to 2/3. What is the correct answer?

6.11 Mr. Smith is the father of two. We meet him walking along the street with a young girl who he proudly introduces as his daughter. What is the probability that Mr. Smith's other child is also a girl? (Assume that boys and girls are equally likely and independent.) Don't just guess. Prove your answer.

6.12 Answer this letter to newspaper columnist Marilyn vos Savant, who is listed in the Guinness Book of World Records Hall of Fame for "Highest IQ":

> *Three of us couples are going to Lava Hot Springs next weekend. We're staying two nights, and we've rented two studios because each holds a maximum of only four people. One couple will get their own studio on Friday, a different couple on Saturday, and one couple will be out of luck. We'll draw straws to see which are the two lucky couples.*
>
> *I told my wife we should just draw once, and the loser would be the couple out of luck both nights. I figure we'll have a two-out-of-three (66 2/3%) chance of winning one of the nights to ourselves. But she contends that we should draw straws twice—first on Friday and then, for the remaining two couples only, on Saturday—reasoning that a one-in-three (33 1/3%) chance for Friday and a one-in-two (50%) chance for Saturday will give us better odds....*
>
> *Which way should we go?*

6.13 Answer this letter to Marilyn Vos Savant: "If I flip a pair of coins until at least one of them lands heads, what are the chances that the other coin also has landed heads?"

Answers

6.1 Most people think that because there are two doors left, the chances are now 50-50. But let's use common sense. We already knew that one of the doors that wasn't chosen had a goat behind it. Does it matter if Monty Hall reminds us that there is a goat behind one of these doors, or if he proves it by showing us a goat? We haven't learned anything useful about the door chosen by the contestant. There is still a 1/3 chance that it is the winning door, and therefore, the probability that the last door is the winner has risen to 2/3. The contestant should switch.

We can also use Bayes' rule or a contingency table with, say, 300 plays of this game, each time with Door 1 being the initial choice. In the 100 cases where the prize is behind Door 1, the host shows Door 2 half the time and Door 3 half the time. When Door 2 or Door 3 has the prize, the host must show the other door. Overall, Door 2 is shown 150 times; Door 3 is shown 150 times. No matter which door is shown, the prize is behind Door 1 1/3 of the time.

	Door 1 Has Prize	Door 2 Has Prize	Door 3 Has Prize	Total
Door 2 shown	50	0	100	150
Door 3 shown	50	100	0	150
Total	100	100	100	300

6.2 As explained in the answer to Exercise 6.1, the probability that the prize is behind door 1 is 1/3, not 1/2.

6.3 The host's opening of a goldfish door did not affect the 1/4 probability that your initial choice has the $10,000 prize. So, if you switch, each of the two remaining doors has a (3/4)/2 = 3/8 probability of being the winning door.

This can be cast as a Bayes' rule problem:

$$P[3 \text{ if show } 1] = \frac{P[3]P[\text{show } 1 \text{ if } 3]}{P[3]P[\text{show } 1 \text{ if } 3] + P[2]P[\text{show } 1 \text{ if } 2] + P[4]P[\text{show } 1 \text{ if } 4]}$$

$$= \frac{(1/4)(1/3)}{(1/4)(1/3) + (1/4)(1/2) + (1/4)(1/2)}$$

$$= 1/4$$

6.4 This is a classic puzzle, known as the Bertrand's box paradox. The answer seems to be 1/2 since the gold coin has ruled out the box with two silver coins, leaving the box with two gold coins and the box with one gold and one silver coins as the only possibilities. However, Bayes' rule gives the correct answer that there is a 2/3 probability that the gold coin came from the box with two gold coins. Let GG, GS, and SS be the three boxes and "G" be the selection of a gold coin:

$$P[GG \text{ if "G"}] = \frac{P[GG]P[\text{"G" if } GG]}{PP[GG]P[\text{"G" if } GG] + P[GS]P[\text{"G" if } GS] + P[SS]P[\text{"G" if } SS]}$$

$$= \frac{(1/3)(1)}{(1/3)(1) + (1/3)(1/2) + (1/3)(0)}$$

$$= 2/3$$

One way to think about it is that there are six, equally likely, coins that could be chosen. The chosen gold coin is equally likely to have come from any of the three gold coins, two of which are from the box with two gold coins.

6.5 The six coins are equally likely to be chosen. Of the three gold coins, two are in the double-gold drawer. So, the probability that you chose the double-gold drawer is 2/3. Therefore, sticking with the drawer you chose has a 2/3 probability of yielding a gold coin.

6.6 There are five equally likely ways for the coin to land heads up and, in three of these cases, the other side of the coin is tails. So, the probability that other side of the coin is tails is 3/5.

6.7 This is like the Monty Hall problem. There are three ways you get a green front side, and in only one of these three cases is the other side red. So, the probability that it is the green-red card is 1/3. Using Bayes' rule:

$$P[GR \text{ if } "G"] = \frac{P[GR]P["G" \text{ if } GR]}{PP[GR]P["G" \text{ if } GR] + P[GG]P["G" \text{ if } GG] + P[RR]P["G" \text{ if } RR]}$$

$$= \frac{(1/3)(1/2)}{(1/3)(1/2) + (1/3)(1) + (1/3)(0)}$$

$$= 1/3$$

Similarly, a contingency table with 300 plays of the game shows that of the 150 times that a green card is seen, it comes from GG 2/3 of the time and GR 1/3 of the time:

	Pick G	Pick R	Total
GG	100	0	100
GR	50	50	100
RR	0	100	100
Total	150	150	300

Therefore the operator will win $4 2/3 of the time and lose $5 1/3 of the time, averaging a $1 profit per game.

6.8 There are three ways to draw a red side and, in two of these cases, the other side of the card is also red. So the probability is 2/3. This can be confirmed with Bayes' rule or a contingency table.

6.9 There are four ways to see a red side: either side of the RR card and the red side of the two RG cards. The probability is consequently one-half that you picked an RG card, in which case, the back side of the card is green.

6.10 This is similar to the Monty Hall problem. A's probability stays at 1/3 because he did not learn anything useful from the warden. This can be confirmed using Bayes' rule.

6.11 Before we learn anything about Mr. Smith's children, there is a .25 probability that he has two boys (BB), a .50 probability that he has one boy and one girl (either BG or GB, depending on birth order), and a .25 probability that he has two girls (GG).

This is like the Monty Hall problem and can be answered by Bayes' rule or a contingency table:

	BB	BG	GB	GG	Total
Observe boy	100	50	50	0	200
Observe girl	0	50	50	100	200
Total	100	100	100	100	400

The probability that Smith's other child is a girl is 100/200 = 1/2.

6.12 Marilyn's (correct) answer is as follows:

> *Actually, it's the same either way. When you draw once for a "loser," as you prefer, you have a 1/3 chance for Friday and a 1/3 chance for Saturday. When you draw twice for winners, as your wife prefers, you have a 1/3 chance for Friday and Saturday alike—not 1/2 for Saturday.*
>
> *This is because you won't be participating in the Saturday drawing at all if you win on Friday, which will be 1/3 of the time. So you only have 2/3 of those 1/2 Saturday chances. And 2/3 of 1/2 equals 1/3—the same as with the other method of drawing.*

An intuitive way to confirm that this answer is correct is to give the three couples names—Couples A, B, and C—and then ask if any of three couples has an advantage over the other two before the drawing starts. If not, then each must have a one-third chance of winning.

6.13 There are two ways that there can be at least one heads on two simultaneous coin flips: two heads or one head and one tail. The latter will happen twice as often as the former so the probability that both coins are heads is 1/3.

7

Binomial Distribution

7.1 A sports columnist for the *Dallas Morning News* had a particularly bad week picking the winners of NFL football games—he got 1 right and 12 wrong, with 1 tie. Afterward, he wrote that, "Theoretically, a baboon at the Dallas Zoo can look at a schedule of 14 NFL games, point to 1 team for each game and come out with at least 7 winners." The next week, Kanda the Great, a gorilla at the Dallas Zoo, made his predictions by selecting pieces of paper from his trainer. Kanda got nine right and four wrong, better than all six *Morning News* sportswriters. Assuming no ties, what is the probability that a baboon picking the winners of 13 games will select at least 9?

7.2 Unanimous jury trials have historically been guaranteed in federal criminal cases, but not in state trials. In Ramos v. Louisiana (2019), the US Supreme Court considered a challenge to the constitutionality of a Louisiana law that allows a conviction if at least 10 of 12 jurors vote guilty. A Stanford law school professor argued that a unanimous verdict of 6 jurors is more trustworthy than a majority decision of 12 or even 20 jurors.

Suppose that each juror is randomly selected from a large pool, of which 90% of the potential jurors would vote guilty and 10% would vote not guilty. Compare the probability that 6 randomly selected jurors would all vote guilty with the probability that a majority of 20 jurors would vote guilty.

7.3 James Fishkin, director of the Center for Deliberative Democracy at Stanford, argued that, "There is so little time for deliberation that some people make leadership choices based on whether they like a candidate's hairstyle." Fishkin proposed that 200–300 randomly selected citizens be asked to spend 1–2 days listening to experts debate a ballot initiative or candidate, and then vote. If 55% of all citizens would vote yes on a ballot initiative after listening to the experts, what is the *exact* probability that more than 50% of 250 randomly selected citizens would vote yes?

7.4 Two baseball teams will play up to five games against each other, with the winner being the first team to win three games. In each game, Team A has a .6 probability of winning and Team B has a .4 probability.

a. What is the probability that Team B will be the winner?

b. Do you think that Team B's chances of being the winner would be higher or lower if it played a series with the winner being the first team to win four games? Explain your reasoning.

DOI: 10.1201/9781003630159-7

7.5 (This is a true story.) On the first day at college, the dean of students said: "Look at the student on your left and the student on your right; one of you won't graduate." Assuming that each student has a 1/3 probability of not graduating, what is the probability that, out of three random elected students, two will graduate and one will not?

7.6 An investment manager wrote that, "If 1,000 people flip coins, some will get heads 10 times in a row, just by chance. You wouldn't anoint them superior coin flippers." If 1,000 people each flip ten coins,

   a. what is the expected value of the number of people who will get ten heads?

   b. what is the probability that at least one person will get ten heads?

7.7 A seemingly healthy woman has a physical checkup which involves a battery of tests of 20 risk factors (such as cholesterol) that might indicate a health problem. For each test, the result is flagged as abnormal if the reading is unusually high or low—specifically, if it is outside a range that encompasses 95% of the readings for healthy women. Thus, if a woman is healthy, there is only a 5% chance that her reading on a test will be outside the normal range. Assuming that the test results are independent, what is the probability that a healthy woman who takes 20 such tests will have two or more abnormal readings?

7.8 In 1980, the United States tried a daring rescue of American hostages that were being held by Iranians in the American Embassy in Teheran. The American plan was to bring in soldiers in eight helicopters traveling 800 miles across the desert at low altitudes during the night. The military commanders believed that at least six of the eight helicopters would have to reach Teheran for the mission to have a chance of succeeding. As it turned out, the rescue attempt was canceled when three of the helicopters became disabled. Assuming independence, calculate the probability that at least six of the eight helicopters would reach Teheran if each has a .75 probability of success. Why might independence not be a warranted assumption?

7.9 Pepys asked Newton which of the following three events is most likely when normal six-sided dice are rolled:

   a. at least one 6 when six dice are rolled.

   b. at least two 6s when twelve dice are rolled.

   c. at least three 6s when eighteen dice are rolled.

   Answer Pepys' question.

7.10 In a 1979 court case, a military supplier argued that the US government had tested too small a sample when it rejected a shipment of 20,000 nose fuse adaptors, a component of artillery shells. Military procurement standards specified that a shipment of 10,001 to 35,000 adaptors would be tested by sampling 315 pieces, yet the government rejected this shipment of 20,000 pieces after testing 20 adapters and finding them all to be defective. If 20 items are randomly selected from an infinite population, in which 10% of the items are defective, what is the probability that all 20 items will be defective? That 15 or more will be defective?

7.11 A basketball article in the *Los Angeles Times* was titled, "Wright Defies the Percentages, and USC Holds Off Oregon, 62–54." The article explained that "Oregon's strategy was obvious in the closing minutes—foul Gerry Wright, a 37.5% free throw shooter. But Wright, a reserve center, didn't fold under the pressure. He made 3 of his 6 foul shots in the final 2 minutes." If a player has a .375 probability of making a shot and each shot is independent, what is the probability of making at least three of six shots?

7.12 Each student admitted to ABC College has a .6 probability of enrolling, and each student's decision is independent of the other students' decisions. Compare a college that admits 1,000 students with a college that admits 2,500 students. Which college has the higher probability that the percentage of students admitted who enroll will be

a. exactly equal to 60%?

b. between 50% and 70%?

c. more than 80%?

7.13 It was reported that a fair coin had been flipped in sixty ten-flip sets (a total of six hundred flips), and that, in two of the ten-flip sets, all ten flips had been heads and that, in another set, all ten flips had been tails. What is the probability that a fairly flipped coin would land either all heads or all tails in three or more of sixty ten-flip sets?

7.14 A marine biologist argued that when there is a long interval between calving, female whales are more likely to have sons than daughters. Data collected by the Center for Coastal Studies found that for whales with an interval of 1–2 years between calving, there were 22 sons and 20 daughters. For mothers calving at intervals of 3, 4, or 5 years, there were 16 sons and only 4 daughters. Assuming that these data are Bernoulli trials with a .5 probability of a son, what is the probability that of 20 calves, 16 or more will be male? (Do not use a normal approximation.)

7.15 In 2018, Eric Reid, a Carolina Panthers football player, said that the National Football League (NFL) was using its purportedly random drug-testing program to target him since he had been tested 7 times in 11 weeks. The NFL and NFL Players Association investigated and concluded that Reid's tests were random. His first test was a mandatory test after he signed with the Panthers. After that initial test, ten players on each NFL team are randomly selected each week. There are 72 players eligible for testing on the Carolina roster; assume that all 32 NFL teams have 72 players eligible for testing. If the testing is truly random, what is the probability that

a. Reid will be selected 6 or more times in 11 weeks?

b. One or more Carolina players will be selected 6 or more times in 11 weeks?

c. One or more teams will have at least one player selected 6 or more times in 11 weeks?

7.16 A researcher tested for precognition by asking a volunteer to predict whether the next color shown on a computer screen will be red or green. The software used a random-event generator with a .5 probability of showing green and a .5 probability of showing red. The volunteer got 60 out of 100 correct. Calculate the exact probability that a volunteer will get at least 60 right merely by guessing. If 25 guessing volunteers are tested, what is the probability that at least one will get 60 or more correct?

7.17 The Silvers are planning their family and want an equal number of boys and girls. Mrs. Silver says that their chances are best if they plan to have two children. Mr. Silver says that they have a better chance of having an equal number of boys and girls if they plan to have ten children. Assuming that boy and girl babies are equally likely and independent of previous births, which probability do you think is higher? Explain your reasoning.

7.18 A state lottery ticket costs $2 and involves picking six different numbers between 1 and 46. The player wins a prize if some or all of the picks match the six randomly chosen numbers drawn by the state. What is the probability of matching *exactly* four numbers? (e.g., selecting the numbers 1, 2, 3, 4, 5, 6 and the winning numbers turning out to be 3, 13, 2, 1, 6, 23.)

7.19 A total of 98 college students were asked how many times they talked to their mothers and fathers during the past year: 36 talked to both parents equally; 9 talked to their fathers more; and 53 talked to their mothers more. If the 62 students who talked more to one parent are equally likely to say mother or father, what is the probability that fewer than 10 talk more to their father?

7.20 Mendel hypothesized that self-fertilization of hybrid yellow-seeded sweet peas would yield offspring with a .75 probability of being yellow-seeded and a .25 probability of being green-seeded. In 1865, he reported that 8,023 such experiments yielded 6,021/8,023 = .7505 yellow-seeded plants and 2,002/8,023 = .2495 green-seeded plants. If these were honest independent trials, with each offspring having a .75 probability of being yellow-seeded, what is the *exact* probability that the number of yellow-seeded plants would be in the range of 6,013–6,021 (fewer than five plants from the expected value, 6,017.25)?

7.21 A study of the anti-parasitic medication Ivermectin for COVID-19 was reported to have involved 600 patients, among whom 410 ages were even numbers and 190 ages were odd numbers. If odd and even ages are equally likely, what is the probability of such a large disparity?

7.22 A renowned stock picker claims that he has better than a 50% chance of picking stocks that will do better than the median stock over the next 12 months. To demonstrate this prowess, he picks ten stocks out of the Russell 3000 (which consists of the 3,000 largest US companies) and a year later it turns out that six have done better than the median return and four have done worse.

    a. If he guesses randomly, what is the probability that six or more of his picks will beat the median return?

    b. If he guesses randomly, would he be more or less likely to get 60% or more correct if he picked 20 stocks instead of 10?

    c. Do your calculations assume that stock returns are normally distributed?

7.23 In a class of 18 students, five 2-person teams will be chosen randomly each week to work on out-of-class projects. (Each week, no one can be on more than one team; each student is either on a team or not on a team.) If there are 10 weeks in the semester, what is the probability that one of the students, Jill McKenzie, will

    a. never be selected?

    b. be selected all 10 weeks?

    c. be selected exactly 5 weeks?

7.24 A statistics professors asked 14 students to imagine 10 coins flip and write down the sequence of heads and tails. Of the 14 students, 13 imagined sequences in which there were 4, 5, or 6 heads. If each student had flipped a fair coin 10 times and written down the sequence that occurred, what is the probability that 13 or more students would have obtained sequences that had 4, 5, or 6 heads?

7.25 Nate Silver's book *The Signal and the Noise* looks at five mortgages, each of which has a 5% chance of defaulting. Assuming independence, he calculates that probability that at least one will default as

$$\binom{5}{1}.05^1.95^4 = .204$$

What is the correct probability?

7.26 A radio station offered a $100 prize to a listener who: (a) has a lucky one dollar bill with at least three 9s among the eight digits in the dollar's serial number (for example, 23944199 and 93944199 would be lucky dollars); and (b) is the correct caller, for example, the 47th person to call the radio station during the contest.

    a. What is the probability that a randomly chosen dollar bill has at least three 9s among the eight digits in the dollar's serial number? (Assume that the digits are independent and that each digit is equally likely to be 0, 1, 2, 3, 4, 5, 6, 7, 8, or 9.)

    b. If you have four dollar bills, what is the probability that at least one will be a lucky dollar?

7.27 The random walk hypothesis says that the probability that stock prices will increase today is independent of past performance. If, on any given day, there is a .52 probability that the Dow Jones Industrial Average will go up and a .48 probability that it will go down, then (assuming independence) what is the probability that during the course of a year with 250 trading days, the Dow will have more up days than down days?

7.28 On average, 70% of the passengers on a flight from Los Angeles to Boston prefer chicken to fish. Assume that these preferences can be modeled as Bernoulli trials. If there are 200 passengers on a flight from LA to Boston and the airline carries 140 chicken dinners and 60 fish dinners, what is the probability that no one will be disappointed? That more than five passengers will be disappointed?

7.29 Jean and Choi agree to flip a coin nine times, with Jean winning if at least five heads appear and Choi winning if at least five tails occur. After four flips, there have been three tails and one head. What is Jean's probability of winning?

7.30 In 1941 Joe DiMaggio hit safely in 56 baseball games. What is the probability that a player who has a .325 probability of getting a hit in each time at bat will have at least one hit in each of the next 56 baseball games? Assume that the Bernoulli trial model applies and that a player has four times at bat in each game. Why might the Bernoulli trial model not be appropriate?

7.31 A coin is to be flipped four times to determine who pays for lunch. You can either predict that heads and tails will each come up twice or you can predict that heads and tails will not come up equally often. Which prediction gives you a higher probability of winning?

7.32 During a basketball game between Chicago and Detroit, the television commentator, Marv Albert, said that Michael Jordan was having trouble at the free-throw line, having made only 9 of 12 shots, while he had made 84% of his free throws going into the game. Assuming the Bernoulli trial model applies, what is the exact probability of 9 or fewer successes in 12 trials, each with a .84 success probability?

7.33 A high school basketball coach who was named "coach of the decade" by the *Los Angeles Times* told his players that missed shots bounce long (away from the shooter) 60% of the time. A study of the video tapes of three men's Division III basketball games found that 84 of 169 missed shots bounced short and 85 bounced long. Assuming that the Bernoulli trial model applies with a probability p that a missed shot will bounce long, determine whether P [X = 85 if p = .6] or P [X = 85 if p = .5] is larger.

7.34 In the board game, Settlers of Catan, players take turns rolling a pair of normal six-sided dice, with every player eligible to receive "resources" based on the number rolled. In one recent game, three of the first six rolls were 3 (a 1 on one die and a 2 on the other die). What are the chances of three or more 3s in the next six rolls? If you play Settlers 500 times, what is the probability that at least one game will begin with three or more 3s on the first six rolls?

Answers

7.1 This is a binomial problem:

$$P[X \geq 9] = \binom{13}{9}.5^9.5^4 + \binom{13}{10}.5^{10}.5^3 + \ldots = .1334.$$

7.2 The probability of a unanimous verdict by six randomly selected jurors is $.9^6 = .531$. The probability that a majority of 20 jurors would vote guilty is given by the binomial distribution:

$$P[X \geq 11] = \binom{20}{11}.9^{11}.1^9 + \binom{20}{12}.9^{12}.1^2 + \ldots + \binom{20}{20}.9^{20}.1^0$$

$$= .999993$$

[Louisiana repealed its law in 2018 (but the repeal was not retroactive), leaving Oregon as the only state that allowed non-unanimous verdicts. In 2020, the Supreme Court decided Ramos v. Louisiana by ruling that unanimous verdicts are required in state criminal cases.]

7.3 The probability is given by the binomial distribution. In general,

$$P[X > 125] = \binom{250}{126}p^{126}(1-p)^{124} + \binom{250}{127}p^{127}(1-p)^{123}$$

$$+ \ldots + \binom{250}{250}p^{250}(1-p)^0$$

Here are some sample calculations

p	P [X > 125]
.40	.000555
.45	.049474
.50	.474794
.55	.936173
.60	.999139

7.4 The fact that they might stop before five games are played is irrelevant. We can simply calculate the probability that one team wins at least three games in a five-game series.

a. Using the binomial distribution:

$$P[X \geq 3] = \binom{5}{3}.4^3.6^2 + \binom{5}{4}.4^4.6^1 + \binom{5}{5}.4^6.6^0 = .317.$$

We can't add together three of five, three of four, and three of three because the teams only play more than three games if neither team wins the first three games. The above calculation is correct because it doesn't matter if the teams were to keep playing after one team has won three games. To check this logic, calculate each team's probability of winning the series using each of these two approaches, and see if the probability of A winning plus the probability of B winning adds to 1.0.

b. The longer the series, the less likely it is that the weaker team will win. With a seven-game series, the probability of B winning drops to .290.

7.5 Using the binomial distribution,

$$P[x = 2] = \binom{3}{2}\left(\frac{2}{3}\right)^2\left(\frac{1}{3}\right)^1 = \frac{4}{9}.$$

7.6 For each person, the probability of ten heads is $(1/2)^{10} = 1/1{,}024$.

a. If we now consider the 1,000 flippers to be $n = 1{,}000$ Bernoulli trials, each with a probability $p = 1/1{,}024$ of successes, then the expected value of the number of people who will get ten heads is $pn = 1{,}000/1{,}024$.

b. The probability that at least one person will get ten heads is equal to one minus the probability that no one gets ten heads:

$$1 - (1 - 1/1.024)^{1{,}000} = .376.$$

7.7 This is a binomial problem:

$$P[x > 1] = \sum_{x=2}^{20} \binom{20}{x} .05^x .95^{20-x} = .264.$$

7.8 The probability is given by the binomial distribution:

$$P[X \geq 6] = \binom{8}{6} .75^6 .25^2 + \binom{8}{7} .75^7 .25^1 + \binom{8}{8} .75^8 .25^0 = .679.$$

The outcomes might not be independent because some conditions (like a severe dust storm) that might disable a helicopter are likely to affect more than one helicopter.

7.9 Using the binomial distribution,

a. $P[x \geq 1] = \sum_{x=1}^{6} \binom{6}{x} \left(\frac{1}{6}\right)^x \left(\frac{5}{6}\right)^{6-x} = .6651.$

b. $P[x \geq 2] = \sum_{x=2}^{12} \binom{12}{x} \left(\frac{1}{6}\right)^x \left(\frac{5}{6}\right)^{12-x} = .6187.$

c. $P[x \geq 3] = \sum_{x=3}^{18} \binom{18}{x} \left(\frac{1}{6}\right)^x \left(\frac{5}{6}\right)^{18-x} = .5973.$

7.10 The probability of 20 defective items in a sample of size 20 can be found from the binomial distribution or from the multiplication rule: $.10^{20} = 1.0 \times 10^{-21}$. The probability of 15 or more is given by the binomial distribution:

$$P[x \geq 15] = \binom{20}{15} .1^{15} .9^5 + \binom{20}{16} .1^{16} .9^4 + \ldots + \binom{20}{20} .1^{20} .9^0$$

$$= 9.48 \times 10^{-12}$$

7.11 This is a binomial problem:

$$P[X \geq 3] = \binom{6}{3} .375^3 .625^3 + \binom{6}{4} .375^4 .625^2$$

$$+ \binom{6}{5} .375^5 .625^1 + \binom{6}{6} .375^6 .625^0 = .404$$

7.12 As the number of trials increases, it is increasingly certain that the success proportion x/n will be close to the success probability p, but less likely that it will be exactly equal to p. Thus the probability is larger for the

a. small college, because the probability that x/n will exactly equal p declines as n increases.

   b.   large college, because the probability that x/n will be close to p increases as n increases.

   c.   small college, because the probability that x/n will be far from p declines as n increases.

7.13 The probability of ten heads and the probability of ten tails are both $(1/2)^{10} = 1/1024$. The probability of one or the other is $2/1024$. The probability that a fairly flipped coin will land either all heads or all tails in three or more of the sixty ten-flip sets is given by the binomial distribution:

$$P[X \geq 3] = \binom{60}{3}\left(\frac{2}{1024}\right)^3\left(\frac{1022}{1024}\right)^{57} + \binom{60}{4}\left(\frac{2}{1024}\right)^4\left(\frac{1022}{1024}\right)^{56} + \ldots$$

$$= .000235.$$

7.14 Using the binomial distribution,

$$P[x \geq 16] = \binom{20}{16}.5^{16}.5^4 + \binom{20}{17}.5^{17}.5^3$$

$$+ \ldots + \binom{20}{20}.5^{20}.5^0 = .0059.$$

7.15 This is a binomial problem with each player having a 10/72 chance of being selected each week. So,

   a.   This is the correct probability calculated by the NFL:

$$P[x \geq 6] = \sum_{x=6}^{11}\binom{11}{x}\left(\frac{10}{72}\right)^x\left(1 - \frac{10}{72}\right)^{11-x} = .001766.$$

   b.   The probability that one or more Carolina players would be selected 6 or more times in 11 weeks is equal to one minus the probability that none of the 72 players would be selected 6 or more times in 11 weeks:

$$1 - (1 - .001766)^{72} = .119.$$

   c.   The probability that one or more teams would have at least 1 player selected 6 or more times in 11 weeks is equal to one minus the probability that no team has one or more players selected 6 or more times in 11 weeks:

$$1 - (1 - .119)^{732} = .982.$$

7.16 Using the binomial distribution, the probability that a guesser will get at least 60 correct is

$$P[x \geq 60] = \sum_{x=60}^{100} \binom{100}{x} .5^x .5^{100-x} = .028444.$$

With 25 guessers, the probability that at least one will get 60 or more correct is equal to one minus the probability that none do:

$$1 - (1 - .028444)^{25} = .514.$$

7.17 Mrs. Silver is correct. This is like coin flips. As the sample size goes up, the chances of exactly 50-50 goes down. (The binomial probabilities are .50 for two children and .2461 for ten children.)

7.18 The probability that the first four numbers chosen will be your selected numbers, followed by two misses is

$$\frac{6}{46} \frac{5}{45} \frac{4}{44} \frac{3}{43} \frac{40}{42} \frac{39}{41}.$$

This is the probability of any single sequence of four hits and two misses. Now we need to count the number of winning sequences, as with the binomial distribution. This is "6 choose 4," so the answer is

$$\binom{6}{4} \frac{6}{46} \frac{5}{45} \frac{4}{44} \frac{3}{43} \frac{40}{42} \frac{39}{41} = .001249$$

or about 1 in 801.

7.19 Using the binomial distribution,

$$P[x < 10] = \sum_{x=0}^{9} \binom{62}{x} .5^x .5^{62-x} = .000000005.$$

7.20 Using the binomial distribution:

$$P[6,013 \leq x \leq 6,021] = \sum_{x=6,013}^{6,021} \binom{8,023}{x} .75^x .25^{8,023-x} = .0924.$$

7.21 Using the binomial distribution with p = .5,

$$P[x \geq 410] = \sum_{x=410}^{600} \binom{600}{x} .5^x .5^{600-x} = 4.9 \times 10^{-20}.$$

The two-sided p-value is 8.18×10^{-20}.

7.22 This is a binomial problem with a .5 success probability.

    a. There is a .5 probability that a randomly picked stock will be above the median, so

$$P[x \geq 6] = \binom{10}{6}.5^6.5^4 + \binom{10}{7}.5^7.5^3 + \ldots + \binom{10}{10}.5^{10}.5^0$$

$$= .377.$$

    b. Less likely if he picks 20 stocks. As the number of trials increases, the probability of getting a result far from the expected value (here, 50%) decreases. The probability of getting 60% or more correct if he picks 20 stocks is .252.

    c. No, the Bernoulli trial assumptions needed for the binomial distribution do not assume normality.

7.23 Each week, Jill has a 10/18 probability of being selected and an 8/18 probability of not being selected.

    a. The probability of never being selected is $(8/18)^{10}$.

    b. The probability of being selected every week is $(10/18)^{10}$.

    c. The probability of being selected exactly five times is given by the binomial distribution:

$$P[x = 5] = \binom{10}{5}\left(\frac{10}{18}\right)^5\left(\frac{8}{18}\right)^5 = .2313.$$

7.24 Using the binomial distribution, the probability of four, five, or six heads is

$$P[x = 4, 5, \text{ or } 6] = \binom{10}{4}.5^4.5^6 + \binom{10}{5}.5^5.5^5 + \binom{10}{6}.5^6.5^4$$

$$= .205078 + .246094 + .205078 = .65625.$$

Again using the binomial distribution, the probability that 13 or more (of 14) students would obtain sequences with 4, 5, or 6 heads is

$$P[x = 13 \text{ or } 14] = \binom{14}{13}.65625^{13}.34375^1 + \binom{14}{14}.65625^{14}.34375^0$$

$$= .020150 + .002748 = .022898.$$

7.25 The probability that at least one will default is equal to one minus the probability that none default:

$$1 - .955 = .2262.$$

7.26 a. This is a binomial problem with $p = .1$, $n = 8$, and $x \geq 3$:

$$P[x \geq 3] = \sum_{x=3}^{8} \binom{8}{x} .1^x .9^{8-x} = .813.$$

b. The probability of at least one lucky dollar among four dollar bills is equal to one minus the probability of no lucky dollars:

$$1 - (1 - .813)^4 = .999.$$

7.27 The answer is given by the binomial distribution:

$$P[x > 125] = \sum_{x=126}^{250} \binom{250}{x} .52^x .48^{250-x} = .716.$$

7.28 The answer is given by the binomial distribution. The probability that no one will be disappointed is

$$P[x = 140] = \binom{200}{140} .7^{140} .3^{60} = .061462.$$

The probability that more than five passengers will be disappointed is equal to the probability that the number of chicken-lovers is less than 135 or more than 145:

$$P[x < 135] + P[x > 145] = \sum_{x=0}^{134} \binom{200}{x} .7^x .3^{200-x} + \sum_{x=146}^{200} \binom{200}{x} .7^x .3^{200-x}$$
$$= .197236 + .198850 = .396086.$$

7.29 Jean has to get at least four heads in next five flips. The binomial distribution gives the probabilities:

$$P[x = 4 \text{ or } 5] = \binom{5}{4} .5^4 .5^1 + \binom{5}{5} .5^5 .5^0$$
$$= .15625 + .03125 = .18750.$$

7.30 The probability of at least one hit in a game is equal to one minus the probability of no hits:

$$1 - (1 - .325)^4 = .7924.$$

The probability of at least 1 hit in the next 56 games is $.7924^{56} = .000002193$. One obvious violation of the Bernoulli assumptions is that the chances of getting a base hit depend on the pitcher.

7.31 The binomial probability distribution with p = .5, n = 4, and x = 2 shows that the probability of two heads is less than .5:

$$P[x = 2] = \binom{4}{2}.5^2.5^2 = .375.$$

7.32 Using the binomial distribution:

$$P[x \le 9] = \sum_{x=0}^{9} \binom{12}{x}.84^x.16^{12-x} = .299.$$

7.33 Using the binomial distribution, it is far less likely that these results came from p = .6 than from p = .5:

$$\frac{P[x = 85 \text{ if } \pi = .6]}{P[x = 85 \text{ if } \pi = .5]} = \frac{\binom{169}{85}.6^{85}.4^{84}}{\binom{169}{85}.5^{85}.5^{84}}$$

$$= \left(\frac{.6}{.5}\right)^{85}\left(\frac{.4}{.5}\right)^{84}$$

$$= \frac{.6}{.5}\left(\frac{.24}{.25}\right)^{84}$$

$$= .0389.$$

7.34 This is a binomial problem with the probability of a 3 on any roll being 2/36. The probability that the next game will begin with three 3s in the first six rolls is

$$P[x \ge 3] = \sum_{x=3}^{6} \binom{6}{x}\left(\frac{2}{36}\right)^x\left(\frac{34}{36}\right)^{6-x} = .003018.$$

For 500 games, the probability that at least one game begins this way is equal to one minus the probability that no game begins this way:

$$1 - (1 - .003018)^{500} = .779.$$

8

Law of Averages

8.1 A sales guru gave an inspirational presentation in which he argued that each *no* brings you closer to a *yes* because 1 out of every 100 phone calls results in a sale. You just have to get through those 99 *nos* to get to a *yes*. What do you say?

8.2 Answer this Ask Marilyn letter: "During the last year, I've gone on a lot of job interviews, but I haven't been able to get a decent job.... Doesn't the law of averages work here?"

8.3 Waldemar Kaempffert, science editor of *The New York Times* for nearly 30 years, dismissed the argument that ESP experiments were mostly lucky and unlucky guesses: "The run-of-luck theory has also been invoked with some disregard of the obvious fact that runs of bad luck will cause runs of good luck in the course of 100,000 trials."

8.4 Explaining why he was driving to a judicial conference in South Dakota, the Chief Justice of the West Virginia State Supreme Court said that, "I've flown a lot in my life. I've used my statistical miles. I don't fly except when there is no viable alternative." What do you suppose the phrase "used my statistical miles" means? Explain why you either agree or disagree with the judge's reasoning.

8.5 After Republicans lost control of the US Senate in the 2020 election, Senator Roy Blunt (R-Mo.) said that, "The optimist in me would say the odds of us getting a break in the future are greater because we've had such a run of bad luck." As a statistician, what do you say?

8.6 After a year of underperformance that was a "complete disappointment," the CEO of an AI-powered fund was optimistic:

> The Horizons Active A.I. Global Equity ETF (MIND), which uses a proprietary AI-directed selection process to invest in major global equity indexes through a basket of ETFs, has trailed the market over the past year, down 1.04% on a return basis, versus a positive total return, in Canadian dollar terms, of 5.76% for the MSCI World Index....
>
> "To the extent that they may have underperformed today, there is a strong possibility that they can outperform in the future," Mr. Hawkins [CEO] said.

Explain Hawkins' reasoning and why you either agree or disagree with it.

DOI: 10.1201/9781003630159-8

8.7 Answer this question to columnist Marilyn vos Savant:

At a lecture on fire safety that I attended, the speaker said: "One in 10 Americans will experience a destructive fire this year. Now, I know that some of you can say you have lived in your homes for 25 years and never had any type of fire. To that I would respond that you have been lucky. But it only means that you are moving not farther away from a fire, but closer to one." Is this last statement correct? Why?

8.8 At the midpoint of the 1991 Cape Cod League baseball season, Chatham was in first with 18 wins, 10 losses, and 1 tie. (Games are sometimes stopped due to darkness or fog.) The Brewster coach, whose team had 14 wins and 14 losses, said his team was in a better position than Chatham to win the championship: "If you're winning right now, you should be worried. Every team goes through slumps and streaks. It's good that we're getting [our slump] out of the way." Do you either agree or disagree with his reasoning?

8.9 Answer this letter to advice columnist Dear Abby:

My husband and I just had our eighth child. Another girl, and I am really one disappointed women. I suppose I should thank God she was healthy, but, Abby, this one was supposed to have been a boy. Even the doctor told me that the law of averages was in our favor 100 to 1.

8.10 As a statistician, how would you answer this March 20, 2024, Yahoo Sports question?

[Shohei] Ohtani went 2-for-5 with an RBI and a walk but he also committed his first [mistake] with the Dodgers, failing to tag second base before heading back to first in the eighth inning. At least he got [his mistake] out of the way early, right?

8.11 [This is a true story.] The father of a Little League baseball player told his son to lend his baseball bat to a poor hitter "to use up the bat's outs." Explain why you agree or disagree.

8.12 [This is a true story.] While a baseball pitcher was making warmup throws before a game, his coach said, "Stop throwing good pitches; you need to get your bad pitches out of the way." What do you think?

8.13 After having lost election contests for Governor of the State of Washington in 2004 and 2008, Dino Rossi ran for the US Senate in 2010. A *Washington Post* columnist wrote, "Is the argument for Dino Rossi basically that, by the law of averages, he has to win something eventually?" As a statistician, how would you respond?

8.14 Casey reasons that if you haven't had a driving accident, your insurance company should raise your rates—not lower them—because you are due for an accident. When he had an accident, he wrote to his insurance company, arguing that everyone has accidents and they should lower his rates now that he has gotten an accident out of the way. Explain why you either agree or disagree with his reasoning.

8.15   Explain why you either agree or disagree with the following:

The golden theorem proven by the Swiss mathematician Jacob Bernoulli states that a variable will revert to a mean in the long run. For example, a small number of flips of one coin might be mostly heads while a small number of flips of another coin might be mostly tails, but, inevitably, a large number of flips will be 50-50 for both coins. In the same way, earnings might be high for one company and low for another, but, inevitably, will be the same for both.

8.16   The law of large number implies that as the number of coin flips increases, it is increasingly likely that the fraction of flips that are heads will be close to 50%. The writer Arthur Koestler once argued that the law of large numbers is wrong because it requires coin flips to be related—a preponderance of heads must be balanced by a preponderance of tails at some point in order for the law of large numbers to be true—yet coin flips are independent, which makes the law of large numbers false. Explain why you either agree or disagree with Koestler's argument.

8.17   Explain why you either agree or disagree with this argument in Leo Gould's book, *You Bet Your Life*:

There are many systems of all kinds but in talking to gamblers you will find the one they believe in most (and it can be used for most gambling games) is the "watch" or "patience" system. It's simple. If you are playing dice and certain bets pay even money, before you put your first [bet] on any of the numbers, watch until one shows up at least three of four consecutive times and then put your first [bet] on the opposite chance.

8.18   A 2013 Bloomberg Businessweek article noted that Warren Buffett's Berkshire Hathaway had underperformed the S&P500 for four months in a row, June, July, August, and September, and concluded that "Berkshire stock is due for a comeback vs. the S&P." Do you agree?

8.19   Explain and evaluate this *Los Angeles Times* writer's appeal to the law of averages:

Thursday night's game was another chapter in the Raiders' stunning success in prime time. Put the Raiders on Monday night, Thursday night, Sunday night football and the only thing in the opponent's favor is the law of averages. Thursday night's win gave the Raiders a 22-2-1 record in prime time.

8.20   Jimmy the Greek, a celebrity gambler, said that,

The professional gambler, when heads comes up four times in a row, will bet that it comes up again....The amateur will figure that heads can't come up again, that tails is "due."

Do you agree with the professional or the amateur? Why?

8.21   Orel Hershiser, before pitching a baseball game for the Los Angels Dodgers, who had not scored a run for 24 consecutive innings, said, "I prefer pitching on the day after we've been shut out because I figure we're going to score. I guess the odds are really in my favor now." Do you agree?

8.22 After the New England Patriots beat the Miami Dolphins in the 1986 National Football League playoffs, a *Los Angeles Times* sportswriter offered this analysis:

> *In the Orange Bowl, though, the Dolphins were basically losing to the law of averages. They had won 18 straight from New England in that stadium.*

Explain and evaluate this use of the law of averages.

8.23 Explain the error in this reasoning: "My next marriage is sure to work; the odds against five divorces are astronomical."

8.24 In his final newspaper column, Melvin Durslag reminisced about what he had learned in his 51 years as a sports columnist, including this advice from a famous gambler:

> *Nick the Greek tipped his secret. He trained himself so that he could stand at the table eight hours at a time without going to the washroom. It was Nick's theory that one in action shouldn't lose the continuity of the dice.*

Explain why you agree or disagree with Nick's secret.

8.25 The "Texas Lotto Guide" gave this advice: "The 'law of probability' essentially means that over a given period of time a number that has not hit [been selected] for a long period of time is due to come in." Do you agree?

Answers

8.1 This is the fallacious law of averages. If you get 99 *no*s, maybe something is wrong with your sales pitch.

8.2 Contrary to the fallacious law of averages, the accumulation of job rejections doesn't increase a person's chances of getting a job offer. In fact, a large number of rejections suggests that this person may have little chance of being offered this type of job.

8.3 He is appealing to the fallacious law of averages.

8.4 The judge evidently believes in the fallacious law of averages.

8.5 This is the law-of-averages fallacy: a run of bad luck does not make good luck more likely.

8.6 This is the fallacious law of averages. In fact, MIND continued to underperform the overall stock market and was closed in 2022.

8.7 This statement relies on the fallacious law of averages; the belief that heads must be offset by tails and good luck must be offset by bad. If anything, 25 years without a fire suggests that this is a careful family with a lower-than-average probability of having a fire.

8.8 He is appealing to the fallacious law of averages—that Chatham's victories must be followed by losses, while Brewster's losses must be followed by victories. By this logic, every team will end up with an equal number of wins and losses. In reality, a team that is in first place at the halfway point is probably better than the other teams, and there is no reason why they should be destined to lose games to offset their wins. As it turned out, Chatham finished the season in first place.

8.9 The doctor may have been figuring the probability of eight girls out of eight babies as roughly 1/100 (actually, $(1/2)^8 = 1/256$). The fallacious law of averages suggests that she is due for a boy, but it is wrong. The odds are still roughly 50-50. (If anything, this history of girl babies suggests that she may be biologically inclined to have girl babies.)

8.10 This is an appeal to the fallacious law of averages.

8.11 This is the fallacious law of averages.

8.12 This sounds like the fallacious law of averages.

8.13 This is the fallacious law of averages. If he keeps losing, the voters evidently don't like him enough. Losing election after election doesn't make someone more likely to win. (Rossi lost the 2010 election but ran for Congress nonetheless in 2018 and lost that election too.)

8.14 This is an example of the fallacious law of averages. Good or bad driving records are statistical evidence of someone being a good or bad driver.

8.15 This sounds like the fallacious law of averages. In addition, there is no reason why these two companies should have the same long-run average earnings.

8.16 The law of large numbers doesn't say that the number of heads and tails must be equal; that is the fallacious law of averages. The law of large numbers says that as the number of flips increases, it is increasingly likely that the fraction that are heads will be close to (not exactly equal to) .50.

8.17 This is the fallacious law of averages.

8.18 This is the fallacious law of averages. Perhaps Buffett lost his Midas touch.

8.19 Past winning doesn't automatically make future wins less likely. If anything, it suggests that the Raiders may be more motivated when they play in prime time.

8.20 The professional believes in hot streaks in coin flips; the amateur believes in the fallacious law of averages. I don't believe in either.

8.21 Hershiser is counting on the fallacious law of averages. Batting poorly one day does not increase the chances of batting well the next day.

8.22   This is the fallacious law of averages—here, that New England losing to Miami so many times made it more likely that New England would win.

8.23   This sounds like the fallacious law of averages. A more reasonable conclusion is that this person is not good at making marriages work, so the next marriage is likely to fail.

8.24   If the dice rolls are independent, there is no "continuity of the dice." Nick should go to the washroom.

8.25   This is the fallacious law of averages. If the Lotto game is fair with each number having an equal chance of being selected, current results do not depend on previous results.

9

Normal Distribution

9.1 A woman wrote Dear Abby, saying she had been pregnant for 310 days before giving birth. Completed pregnancies are normally distributed with a mean of 266 days and a standard deviation of 16 days. What is the probability that a completed pregnancy lasts at least 310 days?

9.2 Daily stock returns are often modeled as random draws from a normal distribution. The daily returns on the S&P 500 index of stock prices over the 50-year period, 1962–2012 had a mean of .031% and a standard deviation of 1.026%. If daily S&P 500 returns are normally distributed with this mean and standard deviation, what is the probability of a daily return as low as −20.3%, which happened on October 19, 1987?

9.3 In his 1868 work, Carl Wunderlich concluded that temperatures above 100.4 degrees Fahrenheit should be considered feverish. In a 1992 study, Maryland researchers suggested that 99.9 degrees Fahrenheit was a more appropriate cutoff. If the oral temperatures of healthy humans are normally distributed with a mean of 98.23 and a standard deviation of .67 (the values estimated by the Maryland researchers), what fraction of these readings is above 100.4? What fraction is above 99.9?

9.4 For a certain microwave oven and popcorn brand, the time taken by a randomly selected popcorn kernel to pop is normally distributed with a mean of 150 seconds and a standard deviation of 20 seconds. What percentage of the kernels do you expect to have popped after 180 seconds?

9.5 A science experiment involving one brand of microwave popcorn found that the time required to pop a kernel is approximately normally distributed with a mean of 129.0 seconds and standard deviation of 9.6 seconds. Assuming that the time needed to pop a kernel is normally distributed with this mean and standard deviation and that kernel pops are independent, how long should we expect it to take for 95% of the kernels to be popped? (Hint: P [Z < 1.645] = .95.)

9.6 Here is histogram of the chest measurements of 5,738 Scottish militiamen in the early 19th century. Based on this figure, make rough estimates of the mean and standard deviation. You must come up with specific numbers and explain how you arrived at them. You will be graded not on whether your estimates are exactly right, but on whether you used a reasonable procedure to obtain your estimates.

DOI: 10.1201/9781003630159-9

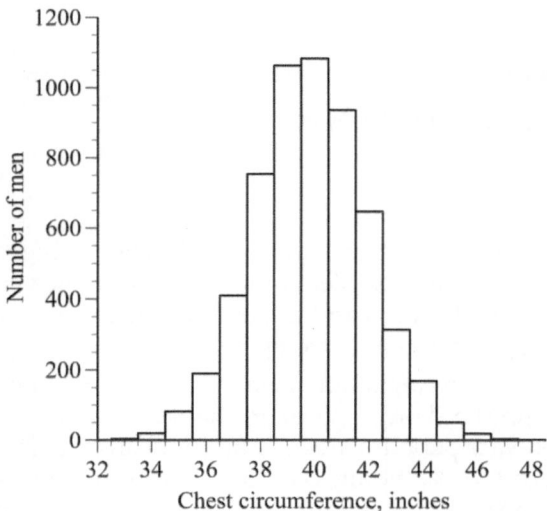

9.7   Over the 141-year period 1878–2018, annual rainfall at the Los Angeles Civic Center averaged 14.50 inches with a standard deviation of 6.68 inches.

     a.   Annual rainfall was 3.60 inches in 2013. If rainfall each year is independent of rainfall in other years and randomly determined from a normal distribution with a mean of 14.50 inches and a standard deviation of 6.68 inches, what is the probability that annual rainfall in a randomly selected year would be so far below average?

     b.   Annual rainfall in Los Angeles is, in fact, not normally distributed. Explain why you believe that it is either skewed left or skewed right.

9.8   A college has found that half of the people they accept for admission decide to enroll. Assuming that enrollment decisions are independent, with each student having a .5 probability of enrolling, how many students should they admit if they want the expected value of the number of students who enroll to be 400? If they follow this admission policy, what is the probability that more than 420 will enroll? Use a normal approximation.

9.9   In the early 1960s, the National Center for Health Statistics estimated the heights of US males between the ages of 25 and 34 to be normally distributed with a mean of 69 inches and a standard deviation of 2.65 inches. Now the mean is estimated to be 70 inches, still with a standard deviation of 2.65 inches. If there are 25 million US males between the ages of 25 and 34, show how to determine the effect of this change in the height distribution on the number of males in this age bracket who are 6 feet 3 inches tall or taller.

9.10 Here is an example of "statistics-based economic analysis."

> Statistics-based economic analysis utilizes the known probabilities associ-
> ated with a bell curve; for example, there is only a 2.5% chance of getting
> an observation that is two standard deviations above the mean. Suppose
> that the unemployment rate over the past 100 months has a mean of 6%
> with a standard deviation of 1%. If the unemployment rate in month 100
> is 8%, there is only 2.5% chance the unemployment rate will go even
> higher, above 8%, in month 101.

What is the logical error in this reasoning?

9.11 A study of 225 female and 100 male residents of a Florida retirement community (all of whom were over the age of 65 with average
ages of 73.7 years for females and 74.1 years for males), found that
cholesterol levels (mg/dl) were approximately normally distributed for each sex with a mean of 228.9 and standard deviation of
37.1 for females and a mean of 202.4 and standard deviation of 36.1
for males. Assuming these to be the population values for elderly
females and males, what is the probability that a randomly selected
elderly female has a cholesterol level above 250. What is the probability that a randomly selected elderly male has a cholesterol level
above 250?

9.12 One measure of the performance of stock portfolio managers is

$$U = \frac{\dfrac{m}{n} - .5}{\sqrt{.5(.5)/n}}$$

where n is the number of years being studied and m is the number of
years that the manager did better than the median manager. Explain in
simple English what $U = -.8$ means.

9.13 You are going to roll m standard six-sided dice n times. Each time you
roll m dice, you will add up the numbers on m dice and record the sum.
When you finish n trials, you will calculate the mean and standard
deviation of n sums and draw a histogram.

a. Which do you expect to have a larger mean, 100 rolls of 4 dice or
1,000 rolls of 4 dice?

b. Which do you expect to have a larger standard deviation, 100
rolls of 4 dice or 1,000 rolls of 4 dice?

c. Which histogram do you expect to have a larger area, 100 rolls of
4 dice or 100 rolls of 40 dice?

d. Which histogram do you expect to look more like a normal distribution, 10,000 rolls of 4 dice or 1,000 rolls of 40 dice?

9.14 Which is more likely to be normally distributed: the number of heads
obtained when 2 coins are flipped 1,000 times or the number of tails
obtained when 10 coins are flipped 100 times?

9.15　Why do you suppose that the time it takes for bags of microwave pop-corn to be fully popped is (approximately) normally distributed, but the time it takes students to answer ten math problems is not?

9.16　The Australian Bureau of Meteorology uses the monthly air-pressure difference between Tahiti and Darwin, Australia, to calculate the Southern Oscillation Index: $SOI = 10(X - \mu)/\sigma$, where X is the air-pressure difference in the current month, μ is the average historical value of X for this month, and σ is the standard deviation of X for this month. Negative values of the SOI indicate an El Niño episode, which is usually accompanied by less-than-usual rainfall over eastern and northern Australia; positive SOI values indicate a La Niña episode, which is usually accompanied by more-than-usual rainfall. Suppose that X is normally distributed with a mean of μ and a standard deviation of σ. Explain why you believe that the probability of an SOI reading as low as –22.8, which occurred in 1994, is closer to 1.1×10^{-15}, .011, or .110.

9.17　The daily high temperatures in two cities on December 31 can be described by normal distributions with City A having a mean of 40° and a standard deviation of 10° and City B having a mean of 50° and a standard deviation of 20°. Which city is more likely to have a daily high temperature below 32° on December 31?

9.18　Suppose that a woman is randomly selected from a population in which heights are normally distributed with a mean of 66 inches and a standard deviation of 2.0 inches and that a male is independently selected from a population in which heights are normally distributed with a mean of 70 inches and a standard deviation of 2.5 inches. For each of the following questions, identify which outcome is more likely (you do NOT have to calculate the probabilities):

　　a.　(i) Her height is more than 70 inches; or (ii) his height is less than 66 inches.

　　b.　(i) Her height is more than 70 inches; or (ii) her height is more than 70 inches and his height is less than 66 inches.

　　c.　(i) Her height is more than 70 inches; or (ii) either her height is more than 70 inches or his height is less than 66 inches.

　　d.　(i) Her height is more than 66 inches and his height is more than 70 inches; or (ii) the sum of their heights is more than 136 inches.

9.19　The median circumference of the waists of adult US males is 34.2 inches. About 2.5% have waists of 29.7 inches or less and 97.5% have waists of 39.5 inches or less. How can you tell from the information given that waist circumferences are not normally distributed? Are waist circumferences positively or negatively skewed?

9.20 Most products are designed for the dimensions of the average person, with a margin that accommodates most but not all people. For instance, the torsos of adult US males are approximately normally distributed with a mean of 32 inches and a standard deviation of 2 inches.

   a. If an automobile is designed with a seat-to-ceiling distance that accommodates a maximum torso length of 35 inches, what percentage of US adult males will not fit comfortably?

   b. A company ordered these cars for all of its employees, of whom 100 are male. If the torso lengths of these males are considered a random selection from the male population as a whole, what is the probability that more than ten males will not fit comfortably? (Use a normal approximation.)

9.21 A *Wall Street Journal* poll in December 1986 asked 35 economic forecasters to predict the interest rate on 30-year Treasury bonds in June 1987. The mean of these 35 predictions was 7.05% and the standard deviation was .53%. If the predictions of all forecasters at that time were normally distributed with a mean of 7.05% and a standard deviation of .53%, what fraction of the forecasters predicted an interest rate of 8.05% or higher, the actual value in June 1987?

9.22 Identify the assumption that is implicit in this claim in the *Financial Analysts Journal*: "Sixty-eight per cent of the time, a portfolio's return will be within plus or minus one standard deviation of its long-term average return."

9.23 Each semester, a teacher shows students a copy of the final exam that was given the previous semester in order to help them study for their final. Once, as the teacher passed out the previous semester's test, he warned, "There is inevitably variation in the difficulty of tests. This test was one standard deviation below the mean." Assuming normality, how would you interpret this remark?

9.24 A variable X is normally distributed with a mean of 10 and a standard deviation of 20. Identify the following statements as either true or false.

   a. $P[X > 15] = P[X < 5]$

   b. $P[X > 15] > P[X > 10]$

   c. $P[X > 20] > P[X < 10]$

9.25 Do you agree or disagree with the following statements about X $\sim N[\mu, \sigma]$?

   a. The median of X is equal to μ.

   b. The standard deviation of $(X - \mu)/\sigma$ is 1.

   c. There is about a .95 probability that X will be more than two standard deviations from μ.

   d. $P[X > \mu + 47] > P[X > \mu]$.

   e. $P[X < \mu - 47] = P[X > \mu + 47]$.

9.26 In 2018 Zillow launched a house-flipping business called Zillow Offers. The idea was that Zillow would use its market-value estimates to make a cash offer to homeowners who want a quick-and-easy sale. Zillow would then use local contractors to make minor repairs and maybe put in a new carpet and a fresh coat of paint and sell the house for a fast profit. The problem turned out to be that homeowners know more about the value of their homes than does Zillow.

   Suppose that the errors in Zillow's price estimates are normally distributed with a mean of zero and a median absolute error of 2%.

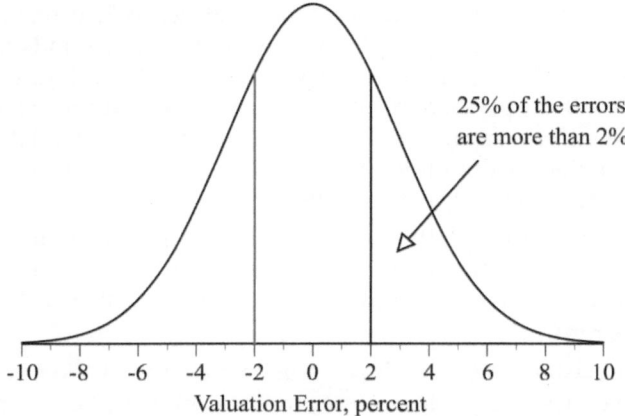

25% of the errors are more than 2%

Valuation Error, percent

   If every seller will only accept an offer from Zillow when the offer is at least 2% too high, make a rough guess of the median valuation error for the homes Zillow buys. Will the mean valuation error be larger or smaller than the median?

Answers

9.1 $Z = (310 - 266)/16 = 2.75$. $P = .00298$.

9.2 Stock returns seem to have "fat tails" with more extreme observations than predicted by the normal distribution:

$$P[X < -20.3] = P\left[\frac{X - \mu}{\sigma} > \frac{-20.3 - .031}{1.026}\right]$$

$$= P[Z < -19.82] = 1.0 \times 10^{-87}$$

9.3 The Z-values are

$$Z = \frac{99.9 - 98..23}{.67} = 2.49$$

$$Z = \frac{100.4 - 98.23}{.67} = 3.24$$

The respective probabilities are P [Z > 2.49] = .0064 and P [Z > 3.24] = .0006.

9.4 The Z-value is Z = (180 – 150)/20 = 1.5. The probability that a kernel will take longer than 3 minutes to pop is P [Z > 1.5] = .0668. So I expect a fraction 1 – .0668 = .9332 (93.32%) to have popped.

9.5 If X = number of seconds to pop 95% of the kernels, then we need 1.645 = Z = (X – 129)/9.6. Solving: X = 129 + 9.6(1.645) = 144.79 seconds.

9.6 The histogram seems symmetrical and roughly bell-shaped with a center at slightly less than 40 inches, so we might use 40 inches for the mean. For the standard deviation, we can use the rule of thumb that for a normal distribution, approximately 95% of the data are within two standard deviations of the mean. It appears roughly that 95% of these data are between 35 and 45. If two standard deviations is equal to 5, then the standard deviation is 2.5 inches.

9.7 a. The Z-value is Z = (3.60 – 14.50)/6.68 = –1.63, or 1.63 standard deviations below average. The probability that rainfall would be this low (or lower) is .052.

b. The rainfall distribution is skewed right because rainfall cannot be negative, but can be far higher than average.

9.8 The probability that an accepted student will attend is p = .5 and, if n students are accepted, then the expected size of the freshman class is pn =. 5n. For this expected size to be 400, n must be 800. The actual number who will attend if 800 are accepted follows the binomial distribution (assuming acceptance decisions are independent of one another). Using a normal approximation,

$$P[X > 420] = P\left[\frac{X - pn}{\sqrt{p(1-p)n}} > \frac{420 - .5(800)}{\sqrt{.5(.5)800}} \right] = P[Z > 1.414] = .0787.$$

9.9 We need to convert to a standardized Z-value in order to find the probability that a randomly selected male in this age bracket will be 75 inches tall or taller:

$$1960s: P[X > 75] = P\left[\frac{X-\mu}{\sigma} > \frac{75-69}{2.65}\right]$$

$$= P[Z > 2.264] = .01179$$

$$2020s: P[X > 75] = P\left[\frac{X-\mu}{\sigma} > \frac{75-70}{2.65}\right]$$

$$= P[Z > 1.887] = .02960.$$

Applying these probabilities to 25 million males, we have .02960 (25,000,000) = 740,000 vs. .01179 (25,000,000) = 294,750.

9.10   When observations are drawn from a normal distribution with a mean of 6 and standard deviation of 1, each selection is independent of previous selections and has a .95 probability of being between 4 and 8. The unemployment rate is different. Next month's unemployment rate is *not* independent of this month's unemployment rate. If the unemployment rate is currently 8%, the unemployment rate next month is far more likely to be near 8% than near 6%.

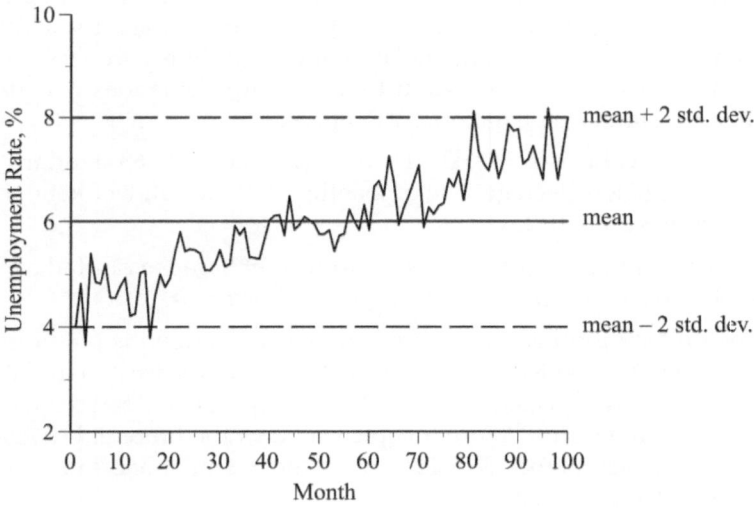

9.11   Converting to standardized Z-values:

$$females: P[X > 250] = P\left[\frac{X-\mu}{\sigma} > \frac{250-228.9}{37.1}\right] = P[Z > .57] = .285$$

$$males: P[X > 250] = P\left[\frac{X-\mu}{\sigma} > \frac{250-202.4}{36.1}\right] = P[Z > 1.32] = .094.$$

9.12 U is the Z-value for a normal approximation to a binomial distribution with p = .5. The value U = − .8 means that this manager is .8 standard deviations below average.

9.13 Here is the theoretical probability distribution for the sum obtained when four dice are rolled. This probability distribution does not depend on how many times the four dice are rolled. The number of times the four dice are rolled only affects how likely it is that the observed histogram is close to the theoretical probability distribution.

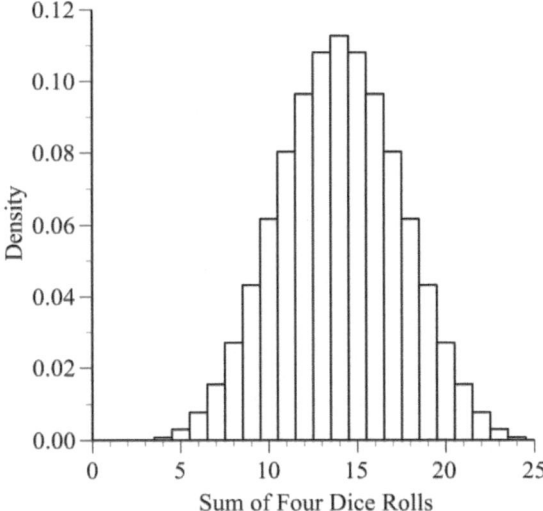

a. Both have the same mean. The expected value for the sum of four dice rolls is 4(3.5) = 14, no matter how many times the four dice are rolled.

b. Both have the same standard deviation. The expected value of the standard deviation for the sum of four dice rolls does not depend on how many times the four dice are rolled.

c. Both have the same area. Histograms always have an area of 1.

d. The histogram of 1,000 rolls of 40 dice will look more like a normal distribution because the central limit theorem depends on the number of dice rolled, not on how many times the experiment is repeated.

9.14 The central limit theorem says that the probability distribution approaches a normal distribution as the number of independent draws being summed increases; so 10 coins flipped 100 times.

9.15 The central limit theorem states that the probability distribution of the sum of n independent, identically distributed random variables approaches the normal distribution as n increases. This might be a reasonable assumption for the factors that determine the popping time of a corn kernel; however, the time it takes a student to do a math problem depends on the student, and the amount of time a student spends on a problem is not independent of the time the student spends on other problems. There might even be a bimodal distribution.

9.16   The SOI is the Z-value multiplied by 10. Therefore, an SOI reading of − 22.8 corresponds to a Z-value of − 2.28, which has a probability of .0113.

9.17   The Z-values are

$$Z_A = \frac{X-\mu}{\sigma} = \frac{32-40}{10} = -.8$$

$$Z_B = \frac{X-\mu}{\sigma} = \frac{32-50}{20} = -.9.$$

City A is more likely to have a daily high temperature below 32°, since a Z-value has a higher probability being below − .8 than being below − .9.

9.18   a.   (ii) because her Z-value is Z = (70 − 66)/2 = 2 and the absolute value of his Z-value is less than 2: Z = (66 − 70)/2.5 = −1.6.

b.   (i) because A is more likely than A and B, unless B is certain.

c.   (ii) because A or B is more likely than A, unless B is impossible.

c.   (ii) because (i) is only one of the many ways that (ii) could be true. Think of it this way: (ii) is both (i) and her height is more than 66 inches, and we know that A is more likely than A and B, unless B is certain.

9.19   A normal distribution is symmetrical, with a normally distributed random variable having a .025 probability of being 2 standard deviations above its mean (which is also its median) and a .025 probability of being 2 standard deviations below its mean. The 2.5% cutoffs given in the exercise are 34.2 − 29.7 = 4.5 inches below the median and 39.5 − 34.2 = 5.3 inches above the median. Because large deviations above the median are more likely than large deviations below the median, the distribution of waist circumferences is positively skewed.

9.20   a.   Converting these data to a Z-value:

$$P[X > 35] = P\left[\frac{X-\mu}{\sigma} > \frac{35-32}{2}\right] = P[Z > 1.5] = .0668.$$

b.   Using the mean and standard deviation of a binomial distribution:

$$P[x > 10] = P\left[\frac{x-\pi n}{\sqrt{\pi(1-\pi)n}} > \frac{10-.0668(100)}{\sqrt{.0668(1-.0668)100}}\right]$$

$$= P[Z > 1.330] = .0918.$$

9.21 Converting these data to a Z-value:

$$P[X > 8.05] = P\left[\frac{X-\mu}{\sigma} > \frac{8.05 - 7.05}{.53}\right] = P[Z > 1.89] = .0294.$$

9.22 This calculation assumes that stock returns are normally distributed.

9.23 Using the normal distribution and one-standard-deviation rule of thumb, there is roughly a .84 probability that a randomly selected test will be more difficult than the test being distributed and a .16 probability that a randomly selected exam test will be less difficult.

9.24 a. True

    b. False

    c. False

9.25 a. Yes

    b. Yes

    c. No (There is about a .95 probability that X will be less than 2 standard deviations from μ.)

    d. No

    e. Yes

9.26 Looking at the region with valuation errors larger than 2%, the median error is not 6% (halfway between 2% and 10%) because there is much more area under the curve between 2% and 6% than between 6% and 10%. The median error has equal areas to the left and right in the range of 2%–10%, and looks to be somewhat lower than 4%. It is, in fact, 3.78%. Since the probability distribution to the right of 2% is skewed right, the mean is higher than the median.

10

One-Sample Tests and Confidence Intervals

10.1 In a letter to the *New England Journal of Medicine*, Dr. Sanders Frank reported that 20 of his male patients with creases in their earlobes had many of the risk factors (such as high cholesterol levels, high blood pressure, and heavy cigarette usage) associated with heart disease. For instance, the average cholesterol level for his patients with earlobe creases was 257 milligrams per 100 milliliters, compared to an average of 215 with a standard deviation of 10 for healthy middle-aged men. If these 20 patients were a random sample from a population with a mean of 215 and a standard deviation of 10, what is the probability that their average cholesterol level would be 257 or higher? Explain why these 20 patients may be a biased sample.

10.2 The random walk hypothesis says that the probability that stock prices will increase today is independent of past performance. If, on any given day, the percent change in the Dow Jones Industrial average is normally distributed with a mean of .1% and a standard deviation of 1.5%, what is the probability that the average of the daily percentage changes over 250 trading days in a year will be positive?

10.3 Explain the error in this Newsweek explanation of the margin of error in a public opinion poll:

> The margin of error, calculated according to a textbook statistical formula, varies inversely with the sample size. In general, about 500 responses give a possible error of 5% either way; 2,500 responses decrease it to 1%.

10.4 A 1995 survey for Fodor's, a publisher of travel books, asked 600 US travelers to name their least desirable vacation destination. The winner (with 9% of the votes) was Iraq/Iran; in second place (with 7%) was New York City. A spokesman for New York's mayor scoffed, "They only interviewed 600 people. That's the size of an apartment building." Assuming this to be a random sample, give a 95% confidence interval for the fraction of all US travelers who would choose New York City as their least desirable vacation destination.

10.5 Suppose that the damage award favored by individual potential jurors in a particular personal injury case can be described by a normal distribution with a mean of $5 million and a standard deviation of

DOI: 10.1201/9781003630159-10

$2 million. (This probability distribution is across randomly selected jurors.)

a. What percentage of potential jurors favor an award of more than $5.5 million? What percentage favor an award of less than $4.5 million?

b. Assume that a jury is a random sample from this distribution and that the jury's damage award is the average of the awards favored by those who serve on the jury. Carefully explain the differences in rewards that can be anticipated with a 6-person jury versus a 12-person jury.

10.6 A certain grade of apple has a weight that is normally distributed with a mean of 10 ounces and a standard deviation of 2 ounces.

a. What is the probability that a randomly selected apple will weigh less than 8 ounces?

b. What is the probability that a bag containing 20 randomly selected apples has a net weight less than 160 ounces?

10.7 The population of Spain is approximately four times the population of Belgium. If random samples are taken of 2,000 Spaniards and 2,000 Belgians, will the margin for error in these polls be (a) four times larger for the Spanish poll; (b) four times larger for the Belgian poll; (c) twice as large for the Spanish poll; (d) twice as large for the Belgian poll; or (e) the same for both polls?

10.8 A July 1987 *Wall Street Journal* poll asked 35 economic forecasters to predict the interest rate on 3-month Treasury bills in June 1988. These 35 forecasts had a mean of 6.19 and a variance of .47. Assuming these to be a random sample, give a 95% confidence interval for the mean prediction of all economic forecasters and then explain why each of these interpretations is correct or not:

a. There is a .95 probability that the actual Treasury-bill rate on June 1988 will be in this interval.

b. Approximately 95% of the predictions of all economic forecasters are in this interval.

c. If the *Journal* took another random sample, there is a .95 probability that the new confidence interval would include 6.19.

10.9 An article about computer forecasting software explained how to gauge the uncertainty in a prediction: "[Calculate] the standard deviation and mean (average) of the data. As a good first guess, your predicted value ... is the mean value, plus or minus one standard deviation—in other words, the uncertainty is about one standard deviation." What did they overlook?

10.10 Does the width of a confidence interval depend on the size of the sample, the size of the population, or the ratio of the sample size to the population size? Explain your reasoning in words not equations.

10.11 Here is an excerpt from a statistical paper:

> *We asked 100 students to guess the number of beans in a glass jar. (There were 183.) Using their guesses, we constructed this 95% confidence interval:*

$$164.25 \pm 1.98\frac{28.71}{\sqrt{100}} = 164.25 \pm 5.70$$

Explain why you either agree or disagree with the following:

a. The calculation should have divided by the square root of 99 instead of the square root of 100.

b. This calculation assumes that the student guesses are normally distributed, which they are not.

c. This confidence interval means that 95% of the student answers were in the interval 164.25 − 5.70 to 164.25 + 5.70.

10.12 Explain the obvious error:

> *The null hypothesis is that the population mean is 0; the alternative hypothesis is that it is not zero. Here, n = 75; X-bar = .004; standard deviation = .0443, t = .781964. Looking at a t table with 74 degrees of freedom, the two-sided p-value is 1.9925.*

10.13 Nationally, 39.6% of all Americans are either first-born or only children. A random-sample of 29 American students at an elite college found that 14 were either first-born or only children.

a. Is this difference from the national average (i) substantial and (ii) statistically significant at the 5% level?

b. How would your answers to the two questions in Part (a) change if the survey had consisted of 290 students of whom 140 were either first-born or only children? Use logic, not equations.

10.14 Identify the error(s):

> *We calculated the fraction p of the soccer players born in the first 6 months of the year by dividing the number born from January to June by the total number of players.*

$$p = \frac{\text{players born from January to June}}{\text{total number of players}}$$

> *Then we set up a one-variable proportion test to figure out whether the fractions born in the first 6 months and last 6 months were equal (p = .5) or not equal (p ≠ .5). We calculated the Z-value by using the equation:*

$$Z = \frac{X - \mu}{p(1-p)/\sqrt{n}}.$$

10.15 Explain why this argument either does or does not make sense to you:

> *The distribution of individual test scores is bi-modal. This means that you cannot do a t-test of the sample mean of all test scores, no matter how large the sample, because that test is based on the assumption that individual scores are normally distributed.*

10.16 An online college statistics class asked this question:

> *A true-false quiz with 10 questions was given to 100 students in a statistics class. Following is the distribution of the scores. Find the mean score and interpret the result. Round the answers to two decimal places as needed.*

Score	5	6	7	8	9	10
Number of Students	5	15	33	28	12	7

What is wrong with the given answer (below)? (Hint: The average score really is 7.48.)

> *The mean score is 7.48. This means that if we were to give this quiz to more and more students, the average score for these students would approach 7.48.*

10.17 The Super Bowl Indicator says that the stock market goes up if a team in the National Football Conference or a team in the American Football Conference that used to be in the National Football League wins the Super Bowl; the market goes down otherwise. This indicator worked for 20 of the first 22 Super Bowls (1967–1988). In 1990 two finance professors published an article in the *Journal of Finance*, a premier peer-reviewed finance journal, arguing that, "although the theoretical relationship connecting the Super Bowl and subsequent stock market movements is not obvious," the 20-out-of-22 record was highly statistically significant and would have been very profitable if followed by investors. To disprove the possibility that it might be a coincidence, they calculated the binomial probability of 20 out of 22 successes to be

$$p[x = 20] = \binom{22}{20}.5^{20}.5^2 = .000055.$$

a. Why is the correct p-value not .000055?

b. What is suspect about the Super Bowl Indicator?

10.18 The life of a Rolling Rock tire is normally distributed with a mean of 40,000 miles and a standard deviation of 5,000 miles. Assuming independence, if you buy four tires, what is the probability that

a. at least 1 tire will last 50,000 miles?

b. all 4 tires will last at least 50,000 miles?

c. the average life will be at least 50,000 miles?

10.19 A 1981 study published in the *New England Journal of Medicine* looked at the use of cigars, pipes, cigarettes, alcohol, tea, and coffee by patients with pancreatic cancer and concluded that there was "a strong association between coffee consumption and pancreatic cancer." This study was immediately criticized on several grounds, including the fact multiple tests were conducted and the most statistically significant result was reported. Subsequent studies failed to confirm an association between coffee drinking and pancreatic cancer. Suppose that independent tests are conducted of six products that are unrelated to pancreatic cancer. What is the probability that at least one of these tests will have a p-value below .05?

10.20 Karl Pearson flipped a coin 24,000 times and obtained 12,012 heads. Use these data to estimate a 95% confidence interval for the probability of heads with this coin. If this coin's probability of a head is .5, what is the exact probability that the number of heads in 24,000 flips would be between 11,988 and 12,012?

10.21 A statistics textbook gives this example:

> Suppose a downtown department store questions 49 downtown shoppers concerning their age.... The sample mean and standard deviation are found to be 40.1 and 8.6, respectively. The store could then estimate μ, the mean age of all downtown shoppers, via a 95% confidence interval as follows:

$$\bar{X} \pm 1.96\frac{s}{\sqrt{n}} = 40.1 \pm 1.96\frac{8.6}{\sqrt{3549}} = 40.1 \pm 2.4.$$

> Thus, the department store should gear its sales to the segment of consumers with average age between 37.7 and 42.5.

Explain why you either agree or disagree with this interpretation: 95% of downtown shoppers are between the age of 37.7 and 42.5.

10.22 A study of 387 corporate stock splits that occurred during the 10 years (2007 through 2016) found that 245 stocks had price increases on the day the split was announced, while 142 had price decreases. Is this difference substantial and statistically persuasive?

10.23 A monthly survey of consumer sentiment was criticized because only 5,000 people were surveyed. In contrast, Twitter tweets can be data mined for millions of people.

     a.   What is the width of a 95% confidence interval for a survey of 5,000 people, if 58% say they are optimistic about the future and 42% say they are not?

     b.   Why might a survey be more reliable than an analysis of Twitter tweets?

10.24 An investor divided her money equally among 100 stocks. She believes that over the next 12 months, each stock's return will be independently determined from a probability distribution with a mean of 10% and a standard deviation of 30%, What can we say about the probability distribution of the return on her portfolio?

10.25 It used to be that, when a National Football League (NFL) game was tied at the end of regulation, the teams played a 15-minute sudden-death overtime with a coin flip used to decide which team kicks off and which team receives. The first team to score wins. During the 2001–2006 seasons, there were 109 sudden-death overtimes, one of which was still tied at the end of the overtime period because neither team scored. In the other 108 games, the team that received the ball first won 64 games and the team that kicked off won 44 games. Use these data to calculate the *exact* two-sided p-value for a test of this null hypothesis: Of those games decided by sudden-death overtime, the kicking and receiving teams have an equal chance of winning. Does the difference seem substantial?

10.26 California birth records were used to obtain data on the ZIP code residence of foreign-born mothers when they gave birth to a daughter and the ZIP code residence of these daughters when they gave birth to a child; 996 of the adult daughters had the same ZIP code as their mothers and 4,298 had different ZIP codes.

a. Estimate a 95% confidence interval for the probability that the adult daughter will have the same ZIP code as her mother.

b. Of the 4,298 daughters who lived in different ZIP codes, 2,579 lived in ZIP codes with higher median income than their mother's ZIP code and 1,719 lived in ZIP codes with lower income. Calculate the exact two-sided p-value for a test of the null hypothesis that a daughter who lives in a different ZIP code is equally likely to live in one with higher or lower income. Is the difference substantial?

10.27 A researcher asked 30 students to read a short article and then estimate how long it took them to read the article. The 30 differences (in seconds) between the actual and estimated times were used to see if people generally overestimated or underestimated. It was reported that a test of the null hypothesis that the population mean is zero gave a t-value of 2.05 and that a 95% confidence interval is 26.53 ± 13.82. How do you know that there was an error in these calculations?

10.28 The first American to win the Nobel prize in physics was Albert Michelson (1852–1931), who was given the award in 1907 for developing and using optical precision instruments. His October 12–November 14,

1882 measurements of the speed of light in air (in kilometers per second) were as follows:

299,883	299,796	299,611	299,781	299,774	299,696
299,748	299,809	299,816	299,682	299,599	299,578
299,820	299,573	299,797	299,723	299,778	299,711
300,051	299,796	299,772	299,748	299,851	

Assuming that these measurements were a random sample from a normal distribution, does a 99% confidence interval include the value 299,710.5 that is now accepted as the speed of light?

10.29 The American Coinage Act of 1792 specified that the gold 10-dollar eagle should contain 247.5 grains of pure gold. An 1837 Act of Congress permitted a .25 grain deviation in the weight of a single eagle and a .048 grain deviation in the average weight of 1000 eagles. If the tested coins are a random sample from a normal distribution with a mean of 247.5 grains and an unknown standard deviation σ, would the Mint be more likely to fail the test of a single coin or the test of 1,000 coins?

10.30 In 1977, the US Supreme Court observed that persons with Spanish surnames constituted 79% of the population of Hildago county, Texas but only 339 of 870 recent jurors in this county. The Court calculated a Z-value of 29 and commented that "as a general rule for such large samples, if the difference between the expected value and the observed number is greater than 2 or 3 standard deviations, then the hypothesis that the jury drawing was random would be suspect to a social scientist."

   a.   What is the null hypothesis?

   b.   Show how the Court calculated its Z-value.

   c.   Explain the reference to "large samples."

   d.   Explain the reference to "2 or 3 standard deviations."

   e.   Does the difference seem substantial to you?

10.31 A high school basketball coach said that a missed free throw by a right-handed player is likely to bounce to the right, while the reverse is true of a left-hander. To test this theory, a researcher asked two college varsity players to shoot 50 free throws. These players missed 21 free throws, of which 15 landed on the same side as their shooting hand, and 6 bounced to the opposite side. Is this difference substantial and statistically persuasive? (Do not use a normal approximation.)

10.32 Sixteen pregnant women were asked to read a children's story aloud three times a day during the last 6.5 weeks of pregnancy. These readings were tape recorded and, shortly after birth, each baby was allowed to choose (by varying the rhythm of their sucking on a special bottle) between a tape recording of their mother reading the familiar story and reading an unfamiliar one. In 13 of 16 cases, the babies chose the familiar story. Is this result substantial and statistically persuasive? (Do not use a normal approximation.)

10.33 In the 1989 and 1990 NCAA basketball tournaments, 1,891 three-point shots were attempted, of which 840 were taken by favored teams and 1,051 by underdog teams. Assuming that these data are a random sample from NCAA tournament games, calculate a 95% confidence interval for the probability that a randomly selected three-point shot was taken by an underdog team.

10.34 Fifty-six college students participated in an ESP test in which a professor flipped a coin ten times and each student attempted to identify each flip as a head or tail.

a. If a student's guess is equally likely to be a head or tail, regardless of his or her previous guesses, what is the probability p that a student will end up with ten guesses consisting of exactly five heads and five tails?

b. It turned out that 27 of these 56 students guessed 5 heads and 5 tails. Use these data to test the null hypothesis that each student has the probability p calculated in Part (a) of guessing exactly five heads and five tails.

10.35 An economics professor told his students that instead of spending hundreds of dollars on a very accurate watch, they should wear 20 cheap watches and calculate the average time. Suppose that the time shown on each cheap watch is normally distributed with a mean equal to the actual time and a standard deviation of 10 seconds, and that the errors on the cheap watches are all independent of each other. What is the mean and standard deviation of the average time on these 20 watches?

10.36 A sociology professor argued that the stress associated with the number 4 might cause Chinese and Japanese persons to suffer an unusually large number of fatal heart attacks on the fourth day of the month. Mortality data for Chinese and Japanese Californians for the years 1989–1998 show that 472 of the 1391 fatal heart attacks that occurred on days 3, 4, or 5 of a month occurred on day 4. Is this substantial and statistically persuasive evidence that more than one-third of the fatal heart attacks that occur on days 3, 4, or 5 of a month happen on day 4? (Use a normal approximation.)

10.37 In 1942, a seed company sold 20 bags of rye seeds with labels indicating a 90% germination rate. The War Food Administration took a sample of seeds from these bags and found that only 240 of the 400 seeds that they tested germinated. Is this difference of any practical importance? Use a normal approximation to calculate a two-sided p-value for testing the null hypothesis that each seed has a .90 probability of germinating.

10.38 Researchers calculated the difference between the actual and predicted daily high temperatures at the Los Angeles Civic Center for every day in 1996. Explain the error(s) in this interpretation of their results. "The t-value for the null hypothesis $\mu = 0$ was 6.47, revealing that there is zero% chance that the population mean is zero. Thus at the 1% level, we disproved the null hypotheses that weather forecasters make no errors in their predictions."

10.39 A test of the null hypothesis that an average college student gains 15 pounds during the first year at college surveyed 100 sophomores and obtained a sample average reported gain of 4.82 pounds with a standard deviation of 5.96 pounds. Explain why you either agree or disagree with each of these conclusions:

a. "When we calculated the t-value, we assumed that the standard deviation of our sample equaled the population standard deviation."

b. "We calculated the t-value to be –17.08. The 2-sided p-value is 2.82×10^{-31}. According to Fisher's rule of thumb, our data are not statistically significant because our p-value is considerably less than 0.05."

c. "Our data strongly indicate that the Freshman 15 is just a myth. However, it must be recognized that we only took one sample of 100 students. Perhaps if we took other samples, our results would be different."

10.40 Americans drive about 3 trillion miles a year and average about 77 reported injuries and 1.09 fatalities per 100 million miles driven. A Google fleet of 55 cars had 11 crashes (2 injuries and no fatalities) while driving 1.3 million miles in autonomous mode between 2009 to 2015.

If a test of autonomous vehicles has no fatalities, how many miles would have to be driven fatality free to reject at the 5% level the null hypothesis that the probability of a fatality is 1 per 100 million miles?

Answers

10.1 The probability is very close to zero because the Z-value is 18.78:

$$Z = \frac{X - \mu}{\sigma\sqrt{n}} = \frac{257 - 215}{10 / \sqrt{20}} = 18.78.$$

Dr. Frank specialized in respiratory medicine. High cholesterol, blood pressure, and cigarette usage are all associated with respiratory problems, so perhaps these patients came to Dr. Frank because cholesterol, blood pressure, and cigarettes were affecting their breathing. Any trait they happen to share (balding, beady eyes, big thumbs) will seem to explain these elevated risk factors, even if they are completely unrelated.

10.2 Converting these data to a Z-value:

$$P[\bar{X} > 0] = P\left[\frac{\bar{X}-\mu}{\sigma/\sqrt{n}} > \frac{0-.1}{1.5\sqrt{250}}\right] = P[Z > -1.054] = .854.$$

10.3 The margin of error varies inversely with the square root of the sample size. To cut the margin of error in half, we need to quadruple the sample size. Here, 500 responses give a margin of error of about 5% (4.47%) and 2,500 responses gives a margin of error of 2%.

10.4 Using a normal approximation to the binomial distribution:

$$p \pm 1.96\sqrt{\frac{p(1-p)}{n}} = .07 \pm 1.96\sqrt{\frac{.07(.93)}{600}} = .07 \pm .02.$$

10.5 a. Converting to a standardized Z-value,

$$P[X > 5.5] = P\left[\frac{X-\mu}{\sigma} > \frac{5.5-5.0}{2.0}\right] = P[Z > .25] = .4013.$$

Similarly, P [x < 4,500,000] is .4013.

b. We can use the sampling distribution for the sample mean to compare 6-person and 12-person juries. If X is normally distributed with mean μ and standard deviation σ, then the sample mean is normally distributed with mean μ and a standard deviation equal to σ, divided by the square root of the sample size.

Thus the awards for both the 6- and 12-person jury systems are normally distributed with the same mean, $5 million. But the standard deviation of the 6-person mean is larger than for the 12-person mean:

$$6-\text{person jury:} \frac{\sigma}{\sqrt{n}} = \frac{2}{\sqrt{6}} = .816$$

$$12-\text{person jury:} \frac{\sigma}{\sqrt{n}} = \frac{2}{\sqrt{12}} = .577.$$

(In general, the standard deviation with a 6-person system is 41.4% larger than with a 12-person system.)

10.6 Transforming to standardized Z-values:

$$P[X < 8] = P\left[\frac{X - \mu}{\sigma} > \frac{8 - 10}{2}\right] = P[Z < -1.00] = .159$$

$$P[\bar{X} < 8] = P\left[\frac{\bar{X} - \mu}{\sigma / \sqrt{n}} < \frac{8 - 10}{2 / \sqrt{20}}\right] = P[Z < -4.47] = .000004.$$

10.7 The correct answer is (e). The margin for error depends on the size of the sample, not on the size of population (unless the sample is a substantial part of the population, which is not the case here).

10.8 A 95% confidence interval for the mean prediction of all forecasters is

$$\bar{X} \pm t_{35-1} \frac{s}{\sqrt{n}} = 6.19 \pm 2.0322 \frac{\sqrt{.47}}{\sqrt{35}} = 6.19 \pm .24.$$

a. Incorrect. This confidence interval is for the average prediction, not the actual value of the T-bill rate.

b. Incorrect. There is a .95 probability that a confidence interval will encompass the true value of the population mean. The distribution of individual forecasts is more dispersed.

c. Incorrect. There is a .95 probability that a new confidence interval will include the true value of the population mean, which almost surely is not 6.19.

10.9 A confidence interval also depends on the sample size.

10.10 The width of a confidence interval depends on the size of the sample since the bigger the sample, the less variation there is in the sample mean.

10.11 a. Disagree. The standard deviation of X requires a division by n − 1 but the standard deviation of a sample mean requires a division by the square root of n.

b. Disagree. This calculation relies on the central limit theorem implying that the sample mean is approximately normally distributed, which does not require the guesses to be normally distributed.

c. Disagree. We expect 95% of such confidence intervals (CIs) to encompass the true value of the population mean. The distribution of the individual student guesses will be much wider than this.

10.12 A p-value cannot be larger than 1.

10.13 a. The sample success proportion, 14/29 = .483, seems substantially higher than .396. An exact binomial test gives a two-sided p-value of 2(.2206) = .4412:

$$p[x \geq 14] = \sum_{x=14}^{29} \binom{29}{x} .396^x .604^{29-x} = .2206.$$

b. If the sample had been ten times larger with the same success proportion, the question of substantial would be the same, but the p-value would be lower:

$$p[x \geq 140] = \sum_{x=140}^{290} \binom{290}{x} .396^x .604^{290-x} = .0017$$

for a two-sided p-value of 2(.0017) = .0034.

10.14 The null hypothesis is that there is a .5 probability that a randomly selected player has a birthdate during the first 6 months of the year. The Z-value should compare the observed frequency, x/n, to the null hypothesis value .5 and the entire denominator should be inside the square root sign and use the null hypothesis value .5, not x/n:

$$z = \frac{x/n - .5}{\sqrt{.5(1-.5)/n}}.$$

10.15 No, the central limit theorem implies that the sample mean approaches a normal distribution even if the individual scores are not normally distributed.

10.16 The number 7.48 is the actual average score of these 100 students; it is not the expected value of an average score.

10.17 a. The p-value should be the probability of 20 or more successes

$$P[x \geq 20] = \binom{22}{20} .5^{20} .5^2 + \binom{22}{21} .5^{21} .5^1 + \binom{22}{22} .5^{22} .5^0 = .000061$$

and it should be a two-sided p-value 2(.000061) = .000122.

b. The calculation of a p-value assumes that the theory was selected before looking at the data. The probability that an extensive search will find some model that is highly correlated with the stock market is 1. The Super Bowl Stock Market Predictor was specified after examining the data, most clearly demonstrated by (a) counting AFC teams that used to be in the NFL as NFC teams; and (b) the words of Leonard Koppett, the man who created the Super Bowl Indicator: "It's a joke! I meant the whole thing as a satire on the fallibility of human statistical reasoning. It's too stupid to believe." Not surprisingly, the Super Bowl Indicator has been mediocre since its creation.

10.18 Writing everything in thousands of miles for convenience, the probability that a randomly selected tire will last 50,000 miles is .0228:

$$P[X > 50] = P\left[\frac{X - \mu}{\sigma} > \frac{50 - 40}{5}\right] = P[Z > 2] = .0228.$$

a. The probability that at least one tire will last 50,000 miles is equal to one minus the probability that none do:

$$1 - (1 - .0228)^4 = .0881.$$

b. The probability that all four tires will last 50,000 miles is $.0228^4 = .000000270$.

c. The probability that the average life of four randomly selected tires will be at least 50,000 miles is

$$P[\bar{X} > 50] = P\left[\frac{\bar{X} - \mu}{\sigma / \sqrt{n}} > \frac{50 - 40}{5 / \sqrt{4}}\right] = P[Z > 4] = .0000317.$$

10.19 The probability of correctly not rejecting the null hypothesis in a single test is $1 - .05 = .95$. The probability of incorrectly rejecting the null hypothesis in at least one case is equal to one minus the probability of not rejecting the null hypothesis in all six cases:

$$1 - .95^6 = .2649.$$

10.20 The fraction that landed heads is $12{,}012/24{,}000 = .5005$. A 95% confidence interval is:

$$p \pm 1.96\sqrt{\frac{p(1 - p)}{n}} = .5005 \pm \sqrt{\frac{.5005(.4995)}{24{,}000}} = .5005 \pm .0063.$$

Using the binomial distribution,

$$P[11{,}988 \le x \le 12{,}012] = \sum_{x=11{,}988}^{12{,}012} \binom{24{,}000}{x} .5^x .5^{24{,}000-x} = .1282.$$

10.21 A 95% confidence interval describes the confidence we have in our estimate of the population mean; it is not a measure of the dispersion in individual ages. These data do not show that 95% (or even a majority) of shoppers are between the ages of 37.7 and 42.5. If ages are normally distributed with a mean of 40.1 and a standard deviation of 8.6, the fraction of shoppers between 37.7 and 42.5 years of age is

$$P[37.7 < X < 42.5] = P\left[\frac{37.7 - 40.1}{8.6} < \frac{X - \mu}{\sigma} < \frac{42.5 - 40.1}{8.6}\right]$$
$$= P[-.279 < Z < .279] = .22.$$

10.22 The proportion 245/387 = .633 seems substantially higher than 50%. Using the binomial distribution,

$$P[X \geq 245] = \binom{387}{245}.5^{245}.5^{387-245} + \binom{387}{246}.5^{246}.5^{387-246} + \ldots$$
$$= .00000009$$

giving a two-sided p-value of .00000018, which is surely statistically persuasive. (The price increases and decreases were actually relative to the overall market on those days; if, for example the market went up 2% and the stock went up 1%, this was counted as a decrease.)

10.23 a.   The width of a confidence interval is just 2 (.0137) = .0274 (2.74% points):

$$\pm 1.96\sqrt{\frac{p(1-p)}{n}} = \pm 1.96\sqrt{\frac{.85(.42)}{5,000}} = \pm .0137.$$

b.   Tweeters are not a random sample and may be very unrepresentative of the population.

10.24 This is like selecting a sample from a probability distribution with a mean of 10 and standard deviation of 30. The probability distribution of the sample mean is approximately normally distributed with a mean of 10% and a standard deviation of 3%:

$$30 / \sqrt{100} = 3.$$

10.25 The exact binomial probability is

$$P[x \geq 64] = \sum_{x=64}^{108}\binom{108}{x}.5^{x}.5^{108-x} = .0335.$$

The two-sided p-value is 2(.0335) = .0670. The receiving team won 59% of the time: 64/108 = .593, which seems substantially higher than 50%. The NFL changed the overtime rules to make it fairer.

10.26 a.   There are a total of 996 + 4,298 = 5,294 daughters, of which 4,298/5,294 = .8119 lived in different ZIP codes. A 95% confidence interval is

$$p \pm 1.96\sqrt{\frac{p(1-p)}{n}} = .8119 \pm 1.96\sqrt{\frac{.8119(.1881)}{5,294}}$$

$$= .8119 \pm .0105.$$

   b.   Under the null hypothesis that the probability is .5, the binomial distribution gives the exact p-value:

$$P[x \geq 2,579] = \sum_{x=2,579}^{4,298} \binom{4,298}{x} .5^x .5^{4,298-x} = 2.98 \times 10^{-40}.$$

   We double this to get the two-sided p-value: 6.0×10^{-40}. The fraction that moved to higher-income ZIP codes seems substantially higher than 50%: 2,579/4,298 = .600.

10.27 The 2.0452 t-value is borderline statistically significant at the 5% level, but the reported 95% confidence interval is far from 0. At least one of these calculations must be wrong. In fact, the correct t-value is 3.93.

10.28 These 23 measurements have a mean of 299,756.22 and a standard deviation of 107.1146. A 99% confidence interval does include 299,710.5:

$$\bar{X} \pm t_{23-1}\frac{s}{\sqrt{n}} = 299,756.22 \pm 2.8190\frac{107.1146}{\sqrt{23}}$$

$$= 299,756.22 \pm 62.96.$$

10.29 The probability that a single coin will fail the test is

$$P[X < 247.5 - .25] = P\left[\frac{X-\mu}{\sigma} < \frac{247.25 - .25 - 247.5}{\sigma}\right]$$

$$= P\left[Z < \frac{-.25}{\sigma}\right].$$

The probability of failing a test of 1,000 coins is

$$P[\bar{X} < 247.5 - .048] = P\left[\frac{\bar{X}-\mu}{\sigma/\sqrt{n}} < \frac{247.25 - .048 - 247.5}{\sigma/\sqrt{1000}}\right]$$

$$= P\left[Z < \frac{-1.52}{\sigma}\right]$$

The 1,000-coin test has a lower probability of failing. (The same is true of the coins being too heavy since the distributions are symmetrical.)

10.30 The Court did one-sample test of a probability.

a. The null hypothesis is that the jurors are a random sample from a population that is 79% Hispanic, so that each juror has a probability p = .79 of being Hispanic.

b. The Z-value is

$$Z = \frac{\frac{x}{n} - p}{\sqrt{\frac{p(1-p)}{n}}} = \frac{\frac{339}{870} - .79}{\sqrt{\frac{.79(1-.79)}{870}}} = -28.99.$$

c. The normal approximation to a binomial distribution is most appropriate for large samples.

d. If the null hypothesis were true, there would be a .0456 or .0026 probability of being 2 or 3 standard deviations from the expected value. Such a low p-value might persuade us to reject the null hypothesis.

e. The fraction of those selected, 339/870 = .390 is substantially different from .79.

10.31 The binomial distribution gives the p-value for testing the null hypothesis that there is a .5 probability of bouncing to same side:

$$P[x \geq 15] = \sum_{x=15}^{21} \binom{21}{x} .5^x .5^{21-x} = .0392.$$

The two-sided p-value is .0784. The fraction that bounced same side, 15/21 = .714, seems substantially larger than .5.

10.32 The success fraction, 13/16 = .8125 seems substantially higher than .5. We can use the binomial distribution to test the null hypothesis that each baby has a .5 probability of choosing the familiar story:

$$P[x \geq 13] = \sum_{x=13}^{16} \binom{16}{x} .5^x .5^{16-x} = .0106.$$

The two-sided p-value is 2(.01060) = .0212.

10.33 The success proportion is 1,051/1,891 = .5558 and a 95% confidence interval is

$$p \pm 1.96 \sqrt{\frac{p(1-p)}{n}} = .5558 \pm 1.96 \sqrt{\frac{.5558(.4442)}{1,891}}$$

$$= .5558 \pm .0224.$$

10.34 a.   The binomial model gives us the probability of exactly five heads and five tails:

$$P[x = 5] = \binom{10}{5}.5^5.5^{10-5} = .2461.$$

b.   If each student has a .2461 probability of guessing 5 heads and 5 tails (H_0: p = .2461), then the probability that 27 or more of 56 students would guess 5 heads and 5 tails is given by the binomial distribution:

$$P[x \geq 27] = \sum_{x=27}^{56} \binom{56}{x}.2461^x(1-.2461)^{56-x} = .00011.$$

10.35 This is a sample mean and it is consequently normally distributed with a mean equal to the actual time and a standard deviation equal to 10 seconds divided by the square root of 20, which is 2.24 seconds.

10.36 The fraction occurring on Day 4 is 472/1,391 = .339, which does not seem substantially different from .333. For testing the null hypothesis that p = 1/3, a normal approximation is

$$P[X \geq 472] = P\left[\frac{X - pn}{\sqrt{p(1-p)n}} \geq \frac{472 - (1/3)1,391}{\sqrt{(1/3)(2/3)1,391}}\right]$$
$$= P[Z \geq .474] = .326.$$

The two-sided p-value is 2(.326) = .652.

10.37 Only 60% of the seeds in the sample germinated, which is substantially below the advertised 90% germination rate. Under the null hypothesis, the p-value is virtually zero:

$$P\left[\frac{x}{n} \leq \frac{240}{400}\right] = P\left[\frac{\frac{x}{n} - p}{\sqrt{p(1-p)/n}} \leq \frac{\frac{240}{400} - .9}{\sqrt{(.9)(.1)/400}}\right]$$
$$= P[Z \leq -20.0] = 2.8 \times 10^{-89}.$$

10.38 Their data are evidence against the null hypothesis that the population mean is zero; that is, the average error is zero. The null hypothesis is not that every error is zero, but that the positive and negative errors balance out, giving an average error of zero.

10.39 a.   They used the standard deviation of the sample to estimate the standard deviation of the population. The use of the t distribution in place of the normal distribution takes into account the fact that this is an estimate, rather than the actual value.

b. With this minuscule p-value, the result is statistically persuasive. The probability of observing this large a difference between the sample mean and the population mean specified by the null hypothesis is virtually 0.

c. The results of another survey would almost certainly be somewhat different; however, the calculated p-value takes into account the sample size. If this was indeed a random sample, it is unlikely that the mean of another sample would be much different and, in particular, not reject the null hypothesis.

10.40 Letting n be the number of miles, the Z statistic is:

$$Z = \frac{\frac{x}{n} - p}{\sqrt{p(1-p)/n}} = \frac{\frac{0}{n} - \frac{1}{100,000,000}}{\sqrt{\frac{1}{100,000,000} \frac{99,999,999}{100,000,000}/n}}.$$

If we set Z equal to the value appropriate for a test at the 5% level, we can solve for n. The solution for n turns out to be very large compared to the number of miles the Google fleet has driven. Setting Z = 1.96:

$$1.96 = \frac{\frac{0}{n} - \frac{1}{100,000,000}}{\sqrt{\frac{1}{100,000,000} \frac{99,999,999}{100,000,000}/n}}$$

$$1.96\sqrt{\frac{1}{100,000,000} \frac{99,999,999}{100,000,000}/n} = -\frac{1}{100,000,000}$$

$$1.96^2 \left(\frac{1}{100,000,000} \frac{99,999,999}{100,000,000}/n\right) = \left(\frac{1}{100,000,000}\right)^2$$

$$n = 1.96^2(99,999,999) = 384,159,996.16.$$

11

Two-Sample Tests and Confidence Intervals

11.1 An A/B test is like an internet laboratory experiment. Two versions of a web page are created—typically the current page and a proposed modification, perhaps with a different banner, headline, or testimonial. A random number generator determines which page a user sees. A pre-specified metric, like purchases or mailing-list signups, is used to determine the winning page.

The Obama 2008 presidential campaign used an A/B test to compare the number of web-page visitors who signed up for more information with these two buttons:

(SIGN UP) (LEARN MORE)

The results:

Button	Visitors	Signups
Sign up	77,858	5,851
Learn More	77,729	6,927

a. Why are the number of visitors not equal?

b. Is the difference between the observed number of signups substantial and statistically persuasive?

11.2 Under its original guidelines, the Environmental Protection Agency (EPA) concluded that second-hand smoke poses a cancer risk if a 95% confidence interval for the difference in the incidence of cancer in the control and treatment groups excluded zero. In 1996 the EPA changed this criterion to a 90% confidence interval. If, in fact, second-hand smoke does not pose a cancer risk, does a switch from a 95% to 90% confidence interval make it more or less likely that the EPA would conclude that second-hand smoke is a cancer risk?

11.3 A French hospital put 412 patients who had suffered one heart attack on a traditional Mediterranean diet (including olive oil, fruit, and bread); the control group consisted of 358 heart-attack patients who were given a recommended low-fat diet. Four of the patients on the Mediterranean diet and 17 of the patients on low-fat diet suffered a second heart attack during the 2 years of the study. Is the observed difference substantial and statistically persuasive?

DOI: 10.1201/9781003630159-11

11.4 A researcher compared the daily returns from a portfolio consisting of stocks that had been deleted from the Dow Jones Industrial Average with a portfolio consisting of the stocks that replaced them. Use these daily return data for a difference-in-means test and a matched-pair test. Which has the lower p-value? Why?

	Deletion Portfolio	Addition Portfolio	Difference in Return
Mean	.000588	.000433	.000155
Standard deviation	.012444	.011862	.008344
Observations	20,367	20,367	20,367

11.5 Evidence from many of sports indicate that performance is enhanced by our personal assessment of our ability. For example, when both contestants in an arm-wrestling study incorrectly believed the weaker person was stronger than his opponent, the weaker person won 10 of 12 matches. When the contestants had correct information about who was stronger, the stronger person won all 12 matches. Is this difference substantial and statistically persuasive?

11.6 A student played 100 games of Roshambo (rock, scissors, paper) against a computer software program, winning 32 games, losing 38, and drawing 30. The computer tried to figure out the player's strategy. This table separates the results into the first 50 games and the last 50 games:

	Player Won	Computer Won	Draw
First 50 games	19	16	15
Second 50 games	13	22	15
Total	32	38	30

a. Is the difference between the first- and second-half win percentages substantial and statistically persuasive?

b. The computer noticed that this player seldom repeated moves (like rock after rock). In 99 opportunities to repeat a move, the player only repeated seven times. Is this a substantial and statistically persuasive pattern?

11.7 A survey of 26 Division III athletes looked at their GPA during the semester they played their varsity sport and during the non-sport semester:

	Sport Season GPA	Off-season GPA	Difference
Mean	3.3250	3.4435	−.1177
Standard deviation	.3385	.3097	.2165

A difference-in-means test gave t = 1.32, with a two-sided p-value of .1939. A matched-pair test gave t = 2.77, with a two-sided p-value of .0104. Explain why these two tests can lead to different conclusions. Which test is more valid?

11.8 In a computer game, Blake is shown (in random order) either: (a) a rectangle that is yellow, blue, green, or red; or (b) the word *yellow, blue, green,* or *red* written in black. Blake's objective is to click as quickly as possible on the word at the bottom of the screen that matches the color or word shown in the middle of the screen. Here are the reaction times in seconds. Is the observed difference in average reaction time substantial and statistically persuasive?

(a) Color	(b) Word	(a) Color	(b) Word
1.12	1.02	1.13	1.07
1.15	1.28	1.13	1.03
.95	.95	.98	.78
1.08	1.17	.95	.88
.93	1.03	1.62	1.05
.82	1.15	1.33	1.20
.88	1.12	.83	1.05
.83	1.18	1.13	1.00
1.17	1.38	1.52	1.20
.90	1.08	.98	1.32

11.9 The Unified Parkinson Disease Rating Scale (UPDRS) gauges the degree of Parkinson's disease in patients on a scale of 0–199, with 0 representing no disability and 199 total disability. Eleven subjects with Parkinson's disease were put on a regiment of riding stationary bikes 3 days a week for 12 weeks. Are the results substantial and statistically persuasive?

Subject ID	UPDRS at Week 0	UPDRS at Week 12
TA	16	3
JC	50	19
LD	23	5
CE	14	9
FE	18	5
SF	48	24
KO	20	3
HP	7	6
DR	2	3
RS	13	0
JT	17	10

11.10 Jerry's time in the 100 freestyle is normally distributed with a mean of 56 seconds and a standard deviation of 2 seconds; Kerry's time is normally distributed with a mean of 57 seconds and a standard deviation of 2.5 seconds. If their swim times are independent, what is the probability that Kerry will beat Jerry?

11.11 A study of 4 years of accident reports by the Dallas Fire Department found that fire trucks painted red made 153,348 runs and had 20 accidents, while fire trucks painted yellow made 135,035 runs and had 4 accidents. Assuming these to be independent random samples, is this difference substantial and statistically persuasive?

11.12 Explain the error(s) in this remark:

> We used a matched-pair test instead of a difference-in-means test because a matched-pair test does not assume that the two samples have the same standard deviation.

11.13 In 1983 22,000 doctors participated in an experiment testing the effects of aspirin on the incidence of heart attacks. Half of the doctors were given a single aspirin tablet every other day; the other half took a placebo every other day. Neither the volunteers or the scientists conducting the experiment knew which volunteers were taking aspirin. After 5 years of what was intended to be a 7-year test, the experiments were stopped by an outside committee monitoring the results. During this period, the doctors taking placebos had 18 fatal heart attacks and the doctors taking aspirin had 5. Is this difference substantial and statistically persuasive? Why do you suppose the experiment was stopped 2 years early?

11.14 In horseshoes, each player pitches two shoes in each inning. If the player pitches two ringers, this is called a double. A "hot hands" study of the 2000 World Horseshoe Pitching championships looked at whether players were more likely or less likely to throw a double after throwing a double in the previous inning. For example, female pitchers threw 1641 doubles and 1691 nondoubles after a nondouble inning, and threw 2142 doubles and 1666 nondoubles after a doubles inning. Is the difference substantial and statistically persuasive?

11.15 A treatment group was given a cold vaccine, while the control group received a placebo. Doctors then recorded the fraction of each group that caught a cold and calculated the two-sided p-value to be .08. Explain why you either agree or disagree with each of these interpretations of these results:

a. "There is an 8% probability that this vaccine works."

b. "If a randomly selected person takes this vaccine, the chances of getting sick fall by about 8%."

c. "These data do not show a statistically significant effect at the 5% level; therefore, we are 95% certain that this vaccine doesn't work."

11.16 What is obviously wrong with this report of the results of a difference-in-means test: "With 38 degrees of freedom, the t-value is 8.98
and the 2-sided p-value is .2916. In this test, the null hypothesis cannot be rejected at the 1% or 5% level because the p-value is far too
high."

11.17 Suppose that you run an experiment involving a treatment group and
a control group with 20 subjects in each sample and a difference-in-
means test gives a t-value of 2.7 and a two-sided p-value of .01. Which
of these statements are true and which are false?

     a. The probability that the null hypothesis is true is .01.

     b. The probability that the treatment works is .99.

     c. If you reject the null hypothesis, the probability that you made
the wrong decision is .01.

     d. If the study is done again, there is a .99 probability of obtaining
the same results.

11.18 A 2021 study compared the calories in 20 McDonald's menu items sold
in the United States and the United Kingdom:

	United States	United Kingdom
Big Mac	550	508
Quarter pounder w/cheese	520	518
Double quarter pounder w/cheese	740	750
Hamburger	250	250
Cheeseburger	300	301
Double cheeseburger	450	445
McChicken	400	388
Filet-o-fish	380	329
Sausage & egg McMuffin	480	430
Six-piece chicken nuggets	255	259
French fries (regular)	320	337
Pancakes	580	477
Apple pie	240	250
McFlurry Oreo	340	258
Chocolate milkshake	520	468
Vanilla milkshake	510	469
Strawberry milkshake	530	458
Cappuccino	120	128
Caramel frappuccino	420	399
Hot chocolate	370	231

    a. Carefully explain how you would determine if the differences are statistically persuasive. You do not need to do any calculations, but you should explain, step-by-step, the calculations you would make.

    b. Carefully explain how you would determine if the observed differences are substantial. Again, do not do any actual calculations.

11.19 Students in one California school district are classified as either a gifted and talented education (GATE) student or a regular student. A survey of seventh- and eighth-graders found that 10 of 86 GATE students had mothers who did not attend college and that 35 of 124 regular students had mothers who did not attend college. Is this difference substantial and statistically persuasive?

11.20 College students rated their intelligence and preparedness for life after college on a scale of 1–10.

    a. Explain why this argument either does or does not make sense.

    I interviewed 150 students of all sexes and years. However, since my study required me to compare two groups (underclassmen and upperclassmen), I couldn't use all my data because there would be more students in one group than the other. So, I randomly selected 10 underclassmen and 10 upperclassmen.

    b. What is wrong with this conclusion?

    I found that despite my original predictions, there was no statistical significance. The p-values were greater than .05, making them statistically insignificant at the 5% level. This means that even though one would expect upperclassmen to feel more intelligent and more prepared for life after college, they do not perceive themselves as so.

11.21 Explain why the proposed test is wrong and identify the correct test:

    My question is "Does playing on home ice in the National Hockey League (NHL) give the home team a better chance to win?" I will choose six NHL teams and, for each team, use a Z single-sample categorical test of the null hypothesis that the chance of winning at home is equal to one half.

11.22 A researcher compared the Internet prices of 20 Abercrombie & Fitch (A&F) male clothing items with the prices of virtually identical clothes sold by Hollister, which is owned by A&F. The average price of the 20 A&F items was $50.13, while the average price of the 20 Hollister items was $31.33. The t-value was 3.38 using a two-sample test and 7.01 using a matched-pair test.

    a. Which test had a lower two-sided p-value?

    b. Explain to someone who has not taken a statistics course why the t-values could be so different.

    c. The matched pair data were used to calculate a 95% confidence interval for the price difference: $18.80 ± 5.62. Explain why you either agree or disagree with this interpretation: "That is to say,

95% of the calculated price difference values will be in the range of $13.18 to $24.42."

11.23 A researcher looked at the 20 English soccer clubs that played in the Premier League during the 2009–2010 season. These teams won an average of 9.65 out of 19 home games, with a standard deviation of 3.79, and won 4.55 out of 19 away games, with a standard deviation of 3.15. (Games that were not won were either lost or tied.) He then calculated the t-value for a two-sample test of the null hypothesis that the average number of wins in home games is equal to the average number of wins in away games:

$$t = \frac{9.65 - 4.55}{\sqrt{\dfrac{3.79}{20} + \dfrac{3.15}{20}}} = 8.82$$

a. What assumption underlying the two-sample t-test is violated?

b. Do you agree that 20 is the correct sample size?

c. What is wrong with this t-value formula?

11.24 California birth records were used to obtain data on the ZIP code residence of mothers when they gave birth to a daughter and the ZIP code residence of these daughters when they gave birth to a child. Of the 4,298 daughters of foreign-born mothers who lived in different ZIP codes, 2,579 lived in ZIP codes with higher median income than their mother's ZIP code and 1,719 lived in ZIP codes with lower income. For the 5,935 daughters of California-born mothers who lived in different ZIP codes, 3,181 lived in ZIP codes with higher median income than their mother's ZIP code and 2,754 lived in ZIP codes with lower median income. Is this difference between the daughters of foreign-born and California-born mothers substantial and statistically persuasive?

11.25 The 794 students in Pomona College's freshmen classes of 1987–1988 and 1988–1989 were separated into two groups: 552 who had attended public high schools and 242 who had attended private schools. Of those from public schools, 48 (8.7%) were Pomona Scholars (an A– grade point average); of those from private schools, 13 (5.4%) were Pomona Scholars. Does the observed difference seem substantial? The author calculated the Z-value to be 1.64 and concluded that although "two classes proved the null hypothesis to be true, …the sample size is minuscule." Did she prove the null hypothesis to be true? Are her results invalidated by the small sample size?

11.26 About one-third of the applicants admitted to one small college actually enroll. Is there a difference between the average SAT scores of those who enroll and those who do not? Here are some recent data:

	Enrolling Students	Non-Enrolling Students
Number of students	368	682
Math score		
Mean	690	700
Standard deviation	63.98	56.44
Reading/Writing score		
Mean	630	640
Standard deviation	72.67	67.54

a. Comparing the enrolling and non-enrolling students, are the observed differences in the math scores substantial and statistically persuasive? What about the reading/writing scores?

b. Explain the error in this interpretation:

According to the central limit theorem, the means of both groups are normally distributed:

$$\bar{X}_1 \text{ is } N\left[690, \frac{63.98}{\sqrt{368}}\right]$$

$$\bar{X}_2 \text{ is } N\left[700, \frac{56.44}{\sqrt{682}}\right]$$

c. Explain the error in this statement:

Under the null hypothesis, the difference in the sample means is zero.

11.27 In the 1980s Kahneman and Tversky reported that when 200 people were asked the following question, 98 answered *yes* and 108 answered *no*:

Imagine that you have decided to see a play and paid the admission price of $10 per ticket. As you enter the theater, you discover that you have lost the ticket. The seat is not marked and the ticket cannot be recovered. Would you pay $10 for another ticket?

When 183 people were asked the following question, 161 answered *yes* and 22 answered *no*:

Imagine that you have decided to see a play where admission is $10 per ticket. As you enter the theater, you discover that you have lost a $10 bill. Would you still pay $10 for a ticket for the play?

Is the difference in responses substantial and statistically persuasive?

11.28 Two polling organizations report these voter preferences:

	John	Jane	Sample Size
Poll A	55%	45%	600
Poll B	52%	48%	200

Can the difference in these poll results be explained by sampling error? If you have no reason to believe one poll is more reliable than the other, what would you use for a prediction of each candidate's vote percentage?

11.29 One hundred randomly selected students were asked to rate the qualifications of two job candidates based on a resume. Half were given the resume for a mythical John Benson and half were given another resume, identical to the first in all respects except that the name was Mary Benson. The results:

	John Benson	Mary Benson
Number of ratings	50	50
Average rating	8.23	8.02
Standard deviation	1.15	2.18

Is the observed difference substantial and statistically persuasive?

11.30 Basketball players are often told that a missed shot usually bounces to the opposite side of the basket from which the shot is taken; for example, that a missed shot from the right side of the basket usually bounces to the left of the basket. A college player suspected that this tendency may be reversed at the end of a game, when the players are tired and their missed shots bounce off the front of the rim. A study of the first and last 10 minutes of videotapes of four different Division III male basketball teams yielded the data shown below. Use these data to see if the difference between the first and last 10 minutes is substantial and statistically persuasive.

	Same Side	Opposite Side
First 10 minutes	16	25
Last 10 minutes	20	10

11.31 Before coming to college, a student watched videotapes of a seminar, "Where there's a will, there's an A," which purports to show scientifically supported strategies for guaranteeing A grades. One strategy is to sit in the front of the class. To test this theory, a statistics student asked 50 randomly selected college students to write down their grade point average (GPA) and to indicate where they typically sit in large classrooms.

	Number of Students	Average GPA (12-Point Scale)
In the very front	5	10.94
Toward the front	21	9.38
In the middle	11	9.38
Toward the back	10	8.37
In the very back	3	10.20

a. The student used a difference-in-means test to compare the average GPA of the 26 students who sit either in the very front or toward the front with the average GPA of all 50 students. What crucial assumption is violated by this procedure? What should he have done instead?

b. After examining the data showing that college students who sit near the front of the classroom have relatively high GPAs, he concluded, "I would certainly advise a freshman to consistently sit in the front row in every single class." Provide a very different interpretation of his data.

11.32 Data were obtained for the finals of the women's 800-meter run at the Southern California Intercollegiate Athletic Conference Track and Field Championships. In three of the years studied (1983, 1986, and 1990), the meet was held on a dirt track; in three other years (1987, 1988, and 1989), an artificial surface was used. Is the observed difference in the average times substantial and statistically persuasive?

	Dirt Track	Artificial
Number of runners	18	18
Average time (seconds)	144.90	143.11
Standard deviation	4.83	4.33

11.33 A study comparing the grades given by female and male students to an essay written by a male found that female graders gave an average grade of 7.125 and male graders gave an average grade of 7.500.

a. A reader said that one flaw in the study was that "the number of male graders was not equal to the number of female graders." Explain why this either is or is not a flaw.

b. Explain why you either agree or disagree with this interpretation of the results: "The t-value was 0.5800, with a p-value of 0.5721. This can be interpreted as there being a 57.21% chance that the means of 7.500 for male graders and 7.125 for female graders were 0.5800 standard deviations away from each other."

11.34 A random sample of 25 male college students and 25 female college students were each asked to name their favorite singer; 21 of the males named a male singer and 16 of the females named a male singer. Explain the error in this test of the null hypothesis that male and female students are equally likely to name a male singer: "The pooled proportion is $(21 + 16)/(25 + 25) = 0.74$. We tested the null hypothesis $p = 0.74$ with the following Z statistic:"

$$Z = \frac{\left(\dfrac{x_1}{n_1} - \dfrac{x_2}{n_2}\right) - 0}{\sqrt{\dfrac{\hat{p}(1-\hat{p})}{n_1} + \dfrac{\hat{p}(1-\hat{p})}{n_2}}}$$

$$= \frac{\dfrac{21}{25} - \dfrac{16}{25}}{\sqrt{\dfrac{.74(1-.74)}{25} + \dfrac{.74(1-.74)}{25}}} = 1.745$$

11.35 From the 30 stocks in the Dow Jones Industrial Average, a researcher identified the ten stocks with the highest dividend-price (D/P) ratios and the ten stocks with the lowest D/P ratios on December 31, 2021. Here are the percentage returns over the next year:

High D/P	Low D/P
39.77	−37.74
32.01	46.83
28.93	−5.81
32.30	48.04
31.28	20.91
12.17	20.24
40.68	1.05
14.33	−28.10
40.39	17.03
37.74	−1.55

a. Display these data in two side-by-side box plots. Do the data appear to have similar or dissimilar means and standard deviations?

b. Is the observed difference in the average returns substantial and statistically persuasive?

11.36 A 2002 study compared radical prostatectomy to watchful waiting for the treatment of localized prostate cancer: 53 of 347 patients in the prostatectomy group died, as did 62 of 348 patients in the control group. The p-value for a test of the null hypothesis that there was no difference in the death rates was .367. Explain why you either agree or disagree with the conclusion of the US Preventive Services Task Force that there was "no difference between the two groups in overall mortality."

Answers

11.1 a. A random number generator was used to determine the page each visitor saw, so we shouldn't expect the number of visitors to be exactly equal.

 b. The signup frequency was 18.6% higher for "Learn More," which seems substantial: $6{,}927/77{,}729 = .0891$ versus $5{,}851/77{,}858 = .0751$. For a difference-in-proportions test, the overall success proportion is

$$\hat{p} = \frac{5{,}851 + 6{,}927}{77{,}858 + 77{,}729} = .0821.$$

 The Z-value is

$$Z = \frac{\dfrac{x_1}{n_1} - \dfrac{x_2}{n_2}}{\sqrt{\dfrac{\hat{p}(1-\hat{p})}{n_1} + \dfrac{\hat{p}(1-\hat{p})}{n_2}}}$$

$$= \frac{.0751 - .0891}{\sqrt{\dfrac{.821(.9179)}{77{,}858} + \dfrac{.0821(.9179)}{77{,}729}}} = -10.03.$$

 The two-sided p-value is 1.097×10^{-23}.

11.2 A 90% confidence interval is narrower than a 95% confidence interval. (That's why it is less likely to include the true value.) Therefore, if the difference in the population means is 0, the difference between the sample means is more likely to be outside a 90% confidence interval.

11.3 The difference in the heart attack rates ($4/412 = .0097$ versus $17/358 = .0475$) seems substantial; those on the low-fat diet were nearly five times more likely to experience a second heart attack. For a statistical test, the pooled proportion is $(4 + 17)/(412 + 358) = .02727$, and the Z-value is

$$Z = \frac{\dfrac{x_1}{n_1} - \dfrac{x_2}{n_2}}{\sqrt{\dfrac{\hat{p}(1-\hat{p})}{n_1} + \dfrac{\hat{p}(1-\hat{p})}{n_2}}}$$

$$= \frac{.0097 - .0475}{\sqrt{\dfrac{.02727(.97273)}{412} + \dfrac{.02727(.97273)}{358}}} = -3.21.$$

The two-sided p-value is .001. (A chi-square test can also be used, and gives the same p-value.)

11.4 The mean return for the Deletion portfolio was 36% higher than the mean return for the Addition portfolio ($.000588/.000433 = 1.36$), which seems substantial. The t-value for a difference-in-means test is

$$t = \frac{\overline{X}_1 - \overline{X}_2}{\sqrt{\dfrac{s_1^2}{n_1} + \dfrac{s_2^2}{n_1}}} = \frac{.000588 - .000433}{\sqrt{\dfrac{.012444^2}{20,367} + \dfrac{.011862^2}{20,367}}} = 1.2867.$$

With 40,638.91 degrees of freedom, the two-sided p-value is .1982.

The t-value for a matched-pair test is

$$t = \frac{\overline{X} - 0}{s/\sqrt{n}} = \frac{.000155 - 0}{.008344/\sqrt{20,367}} = 2.6511.$$

With 20,366 degrees of freedom, the two-sided p-value is .0080.

A difference-in-means test assumes two independent random samples and has to take into account the possibility that, by the luck of the draw, the Deletion portfolio returns might have occurred during a period when the stock market did well and the Addition portfolio returns might have occurred during a period when the stock market did relatively poorly. This possibility is controlled for in a matched-pair test because the time periods are exactly the same.

11.5   The difference seems substantial, 12/12 = 1.0 versus 2/12 = .1667. The overall success proportion is 14/24 = .5833. The Z-value for a difference-in-proportions test is

$$Z = \frac{\dfrac{x_1}{n_1} - \dfrac{x_2}{n_2}}{\sqrt{\dfrac{\hat{p}(1-\hat{p})}{n_1} + \dfrac{\hat{p}(1-\hat{p})}{n_2}}}$$

$$= \frac{\dfrac{12}{12} - \dfrac{2}{12}}{\sqrt{\dfrac{.5833(.4167)}{12} + \dfrac{.5833(.4167)}{12}}} = 4.140.$$

11.6   a.   Omitting the draws, there were 35 winning games in each half. The player won 19 of 35 (54.3%) in the first half and 13 of 35 (37.1%) in the second half, which seems to be a substantial difference. To see whether this difference is statistically persuasive, we can use a difference-in-proportions test:

$$\hat{p} = \frac{19+13}{35+35} = .4571$$

$$Z = \frac{\dfrac{x_1}{n_1} + \dfrac{x_2}{n_2}}{\sqrt{\dfrac{\hat{p}(1-\hat{p})}{n_1} + \dfrac{\hat{p}(1-\hat{p})}{n_2}}}$$

$$= \frac{\dfrac{19}{35} - \dfrac{13}{35}}{\sqrt{\dfrac{.4571(.5429)}{35} + \dfrac{.4571(.5429)}{35}}} = 1.44.$$

The two-sided p-value is .150.

b.   The difference between the player's repeat-move frequency (7/99 = .071) and the 1/3 expected value with random moves seems substantial. We can use the binomial distribution to test the null hypothesis that the player had a 1/3 probability of repeating a move. The probability of seven of fewer repeats is .0000000005:

$$P[x \leq 7] = \sum_{x=0}^{7} \binom{7}{x} (1/3)^x (2/3)^{7-x} = .0000000005.$$

11.7   The difference-in-means test assumes that these are two independent samples. The matched-pair test assumes that observations in each sample are matched with observations in the other sample. Here, we are looking at the sport-season and off-season GPAs for each of these 26 athletes, so a matched-pair test is appropriate. It is also more persuasive because with independent samples we have to consider the possibility that the athletes in one group were better students than those in the other group.

11.8 Here are the summary statistics:

	Sample Size	Sample Mean	Standard Deviation
Color	20	.65	.49
Word	20	.70	.57

The mean reaction time is 7.7% higher for the word than for the color, which may be substantial. A difference-in-means test can be used to determine if the difference is statistically persuasive evidence against the null hypothesis that there is no difference in the population means. Allowing the population variances to differ, the t-value is .298. With 37.2 degrees of freedom, the two-sided p-value is .768. Perhaps a larger sample would have been more persuasive.

$$t = \frac{\overline{X}_1 - \overline{X}_2}{\sqrt{\frac{s_1^2}{n_1} + \frac{s_2^2}{n_1}}} = \frac{.70 - .65}{\sqrt{\frac{.57^2}{20} + \frac{.49^2}{20}}} = .298.$$

11.9 A matched-pair test is appropriate. The t-value for a test of the null hypothesis that the expected value of the improvement is 0 is

$$t = \frac{\overline{X} - 0}{s / \sqrt{n}} = \frac{-12.818}{9.621 / \sqrt{11}} = -4.419.$$

The two-sided p-value is .0013. The 12.82 drop in the average rating from 20.73 to 7.91 is substantial.

11.10 We can use the difference-in-means formula with n = 1:

$$P[X_1 - X_2 < 0] = P\left[\frac{(X_1 - X_2) - (\mu_1 - \mu_2)}{\sqrt{\sigma_1^2 + \sigma_2^2}} < \frac{0 - (57 - 56)}{\sqrt{2.5^2 + 2.0^2}}\right]$$
$$= P[Z < -.3123] = .3774.$$

11.11 The observed accident frequencies are 20/153,348 = .00013 for red trucks and 4/135,035 = .00003 for yellow trucks. Although accidents were infrequent in both cases, they were four times more likely with red trucks, which seems substantial. The overall sample proportion is

$$\hat{p} = \frac{20 + 4}{153,348 + 135,035} = .0000832.$$

The Z-value is

$$Z = \frac{\dfrac{20}{153,348} - \dfrac{4}{135,035}}{\sqrt{\dfrac{.0000832(1-.0000832)}{153,348} + \dfrac{.0000832(1-.0000832)}{135,035}}}$$

$$= 2.96.$$

The two-sided p-value is .0031, which seems statistically persuasive. Many fire departments no longer use traditional red trucks.

11.12 A difference-in-means test does not have to assume that the two samples have the same standard deviation. A matched-pair test does require genuinely matched pairs.

11.13 The observed fatality frequencies are 18/11,000 = .00013 for placebo and 5/11,000 = .00003 for aspirin. Although fatalities were infrequent in both cases, they were 3.6 times more likely with the placebo, which seems substantial. The overall sample proportion is

$$\hat{p} = \frac{18+5}{11,000+11,000} = .0010.$$

The Z-value is

$$Z = \frac{\dfrac{18}{11,000} - \dfrac{5}{11,000}}{\sqrt{\dfrac{.0010(1-.0010)}{11,000} + \dfrac{.0010(1-.0010)}{11,000}}}$$

$$= 2.71.$$

The two-sided p-value is .0067, which seems statistically persuasive (and persuaded the researchers to stop the test 2 years early). Many doctors now recommend aspirin for at-risk patients.

11.14 The difference in success proportions (2,142/3,808 = .5625 versus 1,641/3,332 = .4925) is not overwhelming but can be the difference between winning and losing a match. The overall success proportion is

$$\hat{p} = \frac{1,641+2,142}{3,332+3,808} = .5298.$$

A difference-in-proportions test gives:

$$Z = \frac{\dfrac{2,142}{3,808} - \dfrac{1,641}{3,332}}{\sqrt{\dfrac{.5298(1-.5298)}{3,808} + \dfrac{.5298(1-.5298)}{3,332}}} = 5.91.$$

This difference is surely statistically persuasive, with p < .000001. (A chi-square test gives chi-square = 34.96, which (except for rounding) is equal to the Z-value squared. The p-value is the same.)

11.15 a.   False. The p-value is the probability that the sample frequency would be so far from the null-hypothesis, not a statement about the probability that vaccine works. Indeed, classical statisticians would not put a probability on a null hypothesis being true or false.

  b.   False. A p-value doesn't measure the size of an effect.

  c.   False. See (a).

11.16 There is a mistake in either the reported t-value or p-value because a t-value of 8.98 with 38 degrees of freedom would have a minuscule p-value.

11.17 These statements are all false.

11.18 a.   We can do a matched-pair test of the null hypothesis that the average difference is equal to zero. The steps are to calculate: the difference between each pair of calories; the mean and standard deviation of these differences; the t-value for a one-sample test

$$t = \frac{\overline{X} - 0}{s / \sqrt{n}}.$$

and the two-sided p-value with $20 - 1 = 19$ degrees of freedom. The answers are shown below:

	United States	United Kingdom	United States – United Kingdom
Mean	413.75	382.65	32.1
SD	145.5740	139.5890	42.0925

$$t = \frac{\overline{X} - 0}{s / \sqrt{n}} = \frac{32.1 - 0}{42.0925 / \sqrt{20}} = 3.41$$
$$2p = .0029.$$

  b.   To determine if the differences are substantial, I would compare the average difference to the average level (or perhaps look at the average percent difference). Here, the average difference between the US and UK calories is 32.1, which is 8.4% higher than the average UK calories. Some may think this is substantial; others may disagree.

11.19 The difference in the fraction of mothers who did not attend college seems substantial: $35/124 = .282$ versus $10/86 = .116$. For a statistical test, the overall success proportion is

$$\hat{p} = \frac{35 + 10}{124 + 86} = .2143.$$

and the Z-value is

$$Z = \frac{\dfrac{35}{124} - \dfrac{10}{86}}{\sqrt{\dfrac{.2143(1-2143)}{124} + \dfrac{.2143(1-2143)}{86}}} = 2.88.$$

The two-sided p-value is .0039, which seems persuasive.

11.20 a. There is no compelling reason why the sample sizes have to be equal. The statistical test will take into account unequal sample sizes.

b. A failure to reject the null hypothesis does not prove that the null hypothesis is true.

11.21 Why should we assume that every team has a 50% chance of winning its home games? This should be difference-in-proportions test of the null hypothesis that a team is equally likely to win at home or on the road. (A chi-square test with a 2 × 2 table would also work.)

11.22 a. The matched-pair test had a lower two-sided p-value since it had the higher t-value.

b. A matched-pair test controls for the possibility that, with two independent samples, one sample might include items that are more expensive because of what they are rather than which store they are sold in.

c. The correct interpretation of a 95% confidence interval is that there is a 95% probability that a confidence interval calculated in this way will include the value of the population mean.

11.23 a. The test assumes that the samples are independent, but home and away wins are not independent since if a team wins a home game, the opponent loses an away game.

b. No, the test should take into account the number of games played.

c. The standard deviations should be squared.

11.24 The difference in the proportion who moved to higher-income ZIP codes seems substantial (though some may disagree): 2,579/4,298 = .600 for foreign-born mothers and 3,181/5,935 = .536 for California-born mothers. Using a difference-in-means test with a common estimate of the success probability,

$$\hat{p} = \frac{2,579 + 3,181}{4,298 + 5,935} = .5629$$

$$Z = \frac{\dfrac{2,579}{4,298} - \dfrac{3,181}{5,935}}{\sqrt{\dfrac{.5629(1-.5629)}{4,298} + \dfrac{.5629(1-.5629)}{5,935}}} = 6.45.$$

The two-sided p-value is .0000000001. This exercise can also be answered with a chi-square test with 1 degree of freedom, which gives a chi-square value of 41.5935, and is equal to the square of the Z-value for a difference-in-means test. The p-values are the same.

11.25 The difference between 8.7% and 5.4% seems substantial. Not rejecting the null hypothesis does not prove the null hypothesis to be true. The small sample size is taken into account in the standard deviation calculation for the Z-value and does not invalidate her results.

11.26 a. The ten-point differences in each average do not seem substantial. However, the differences are statistically persuasive (at least by Fisher's 5% rule), as the Z-values (allowing possibly unequal standard deviations) are

$$\text{math: } t = \frac{700 - 690}{\sqrt{\dfrac{56.44^2}{682} + \dfrac{63.98^2}{368}}} = 2.52$$

$$\text{reading/writing: } t = \frac{640 - 630}{\sqrt{\dfrac{67.544^2}{682} + \dfrac{72.67^2}{368}}} = 2.18.$$

For math, the two-sided p-value with 675.71 degrees of freedom is .0121; for reading/writing, the two-sided p-value with 706.17 degrees of freedom is .0296.

b. According to the central limit theorem, the sample means are normally distributed with unknown values of the population mean and standard deviations, not the sample mean and standard deviation.

c. Under the null hypothesis, the difference in the population means is zero. The observed difference in the sample means will almost surely not equal zero.

11.27 The difference between the sample success proportions seems substantial: 92/200 = .4600 and 161/183 = .8798. For a statistical test, we can use the pooled sample success proportion, (92 + 161)/(200 + 183) = .6606 to calculate the Z-value:

$$Z = \frac{\dfrac{161}{183} - \dfrac{92}{200}}{\sqrt{\dfrac{.6606(1 - .6606)}{183} + \dfrac{.6606(1 - .6606)}{200}}} = 8.67.$$

The two-sided p-value is minuscule.

11.28 Using the overall success proportion,

$$\hat{p} = \frac{.55(600) + .52(1200)}{600 + 1200} = .53.$$

The Z-value for a difference-in-proportions test is

$$Z = \frac{.55 - .52}{\sqrt{\dfrac{.53(1-.53)}{600} + \dfrac{.53(1-.53)}{1200}}} = 1.202.$$

The two-sided p-value is .229, which means that if these are random samples from the same population, there is a .229 probability that sampling error will yield a difference in poll results as large as or larger than the difference observed here. I would combine the two poll results into a single poll with 1800 observations, 53% preferring John and 47% preferring Jane.

11.29 The average rating for John was 2.6% higher than the average rating for Mary. It is debatable whether this is substantial. The t-value using possibly different standard deviations is

$$t = \frac{8.23 - 8.02}{\sqrt{\dfrac{1.15^2}{50} + \dfrac{2.18^2}{50}}} = .6025.$$

The two-sided p-value with 74.31 degrees of freedom is .549, which does not seem statistically persuasive.

11.30 The difference in the same-side success proportions seems substantial: 20/30 = .667 versus 16/41 = .390. For a statistical test, the overall success proportion is

$$\hat{p} = \frac{20 + 16}{30 + 41} = .507$$

$$z = \frac{\dfrac{x_1}{n_1} - \dfrac{x_2}{n_2}}{\sqrt{\dfrac{\hat{p}(1-\hat{p})}{n_1} + \dfrac{\hat{p}(1-\hat{p})}{n_2}}}$$

$$= \frac{\dfrac{20}{30} - \dfrac{16}{41}}{\sqrt{\dfrac{.507(1-.507)}{30} + \dfrac{.507(1-.507)}{41}}} = 2.301.$$

The two-sided p-value is .0214, which is statistically persuasive by Fisher's 5% rule. (A chi-square test can also be done and yields the same results.)

11.31 a. A difference-in-means test is supposed to use two independent random samples. This is not the case here because the 26 students in the front group are also included in the total sample of 50 students. He should have compared the 26 in front with the 13 in back or with the remaining 24.

b.　The correlation does not prove that sitting in front causes people to get higher grades. The causation may be the other way around: perhaps the most serious students sit in the front so that they can hear the professor clearly and read the whiteboard, while those who want to daydream or are unprepared to answer questions try to hide farther back.

11.32 Using a difference-in-means t-test with possibly unequal standard deviations,

$$t = \frac{144.90 - 143.11}{\sqrt{\dfrac{4.83^2}{18} + \dfrac{4.33^2}{18}}} = 1.171.$$

With 33.60 degrees of freedom, the two-sided p-value is .2498, which does not seem statistically persuasive. It is a matter of opinion whether the 1.79-seconds difference in average times is substantial.

11.33 a.　A difference-in-means test does not require equal sample sizes.

b.　The p-value is the probability of observing a difference in the sample means as large as or larger than that observed if the population means are equal.

11.34 Their Z-value is correct. However, they are not testing the null hypothesis that $p = .74$; they are testing the null hypothesis that $p_1 = p_2$.

11.35 a.　The box plots show the high D/P stocks to generally have higher returns than the low D/P stocks and to have considerably less variation in the returns (we shouldn't assume the population standard deviations are equal):

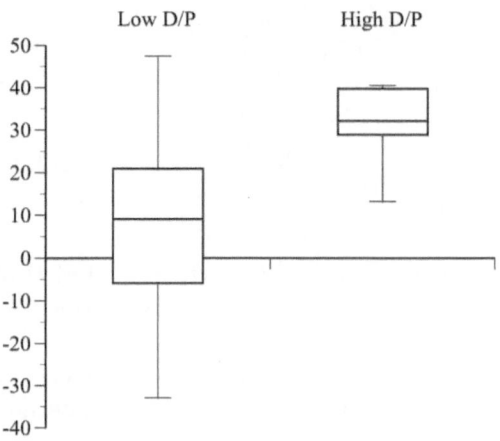

b. The sample means and standard deviations confirm the visual impression in the box plots. The high D/P stocks did much better, on average, than the low D/P stocks, and with considerably less dispersion in the returns:

	High D/P	Low D/P
Mean	30.96	8.09
Standard deviation	10.24	28.35

The t-value for a difference-in-means test is 2.399:

$$t = \frac{30.96 - 8.09}{\sqrt{\dfrac{10.24^2}{10} + \dfrac{28.35^2}{10}}} = 2.399.$$

Statistical software shows that there are 11.31 degrees of freedom and the two-sided p-value is .0347. The observed differences in the mean returns is substantial and barely statistically significant at the 5% level.

11.36 The p-value does not measure whether the difference is substantial and meaningful. The death rate fell by 14%: (53/347)/(62/348) = .8573. Although the difference in overall death rates is not statistically persuasive, the 14% drop in death rates seems substantial.

12

Chi-Square Tests

12.1 It has been argued that when people are asked to choose a number, say from 1 to 4 or 1 to 10, they tend to choose an "average number" that is in the middle rather than an "extreme number" on either end. To test this theory, a professor did a mock ESP experiment, in which 40 students were asked to guess which one of four cards, numbered 1–4, the professor had chosen randomly. Are the patterns in these data substantial and statistically persuasive?

Number Guessed	Number of Students
1	2
2	13
3	21
4	4

12.2 In the game Roshambo (rock-scissors-paper), two players simultaneously show a fist (rock), two fingers (scissors), or an open hand (paper). Rock beats scissors, scissors beat paper, and paper beats rock. A researcher played this game against 120 people, recording the initial move of each opponent. Use these data to test the null hypothesis that rock, scissors, and paper are used equally often on the initial move: 27 rock, 54 scissors, and 39 paper. Are the observed differences substantial?

12.3 When the Titanic sank in 1912, it was carrying 1,315 passengers: 402 adult females, 805 adult males, and 108 children. Of the 498 passengers who were saved, there were 296 adult females, 146 adult males, and 56 children. In which group was the fraction saved disproportionately large? Disproportionately small? Do these data reject at the 1% level the null hypothesis that a person's chances of being saved did not depend on the group he or she was in?

12.4 A statistics professor flipped a fair coin 10 times and asked each of the 24 students to guess whether the flip was a head or tail. Each student then found the longest streak in their guesses; for example, in these data H T H H T H T T T H, the longest streak is 3. The results were:

DOI: 10.1201/9781003630159-12

Longest Streak	Number of Students	True Probability of a Streak This Long
< 3	4	.174
3	17	.361
> 3	3	.465

What pattern do you see in these results? Are the deviations from the expected values substantial and statistically persuasive? If so, how would you explain these patterns?

12.5 A study of major league baseball (MLB) managers from 1871 through 2007 found that 540 managers played professional baseball before becoming a manager, with the primary positions grouped as follows: pitcher (51), catcher (116), infielder (262), and outfielder (111). Use these data to test the null hypothesis that the respective probabilities are 1/9 pitcher, 1/9 catcher, 4/9 infielder, and 3/9 outfielder. What patterns do you see in your data? Are these patterns substantial and statistically persuasive?

12.6 The five positions on a basketball team are shown below. As of 2012, 167 of the 299 head coaches in the NBA played in the NBA. Their playing positions:

	Number
Point guard	61
Shooting guard	42
Small forward	25
Power forward	21
Center	18
	167

Is there substantial and statistically persuasive evidence against the hypothesis that coaches are equally likely to come from each of these five positions?

12.7 A study looked at the positions played by the 70 managers in the top five professional soccer leagues during the 2022 season who had played professional soccer:

Forward	14
Midfielder	36
Defense	20
Goalie	0

Assuming that 3/11 of all players are forwards, 3/11 midfielders, 4/11 defenders, and 1/11 goalies, what patterns do you see in your data? Are these patterns substantial and statistically persuasive?

12.8 When 100 MIT MBA students were offered these three subscription choices to *The Economist*, 16 chose the online subscription, none chose the print subscription, and 84 chose the combination subscription:

1-year online subscription	$59
1-year print subscription	$125
1-year print and online subscription	$125

When 100 other students were offered just two choices, 68 chose the online subscription and 32 chose the combination subscription:

1-year online subscription	$59
1-year print and online subscription	$125

Is this observed difference in the online and combination subscription choices between these two offers substantial and statistically persuasive?

12.9 A student asked a random sample of 54 of her classmates whether they were nearsighted and whether their grade point average (GPA) was above or below B+:

	Nearsighted	Not Nearsighted
GPA < B+	14	15
GPA ≥ B+	17	8

a. Are these data statistically persuasive evidence that nearsightedness and GPA are related?

b. This student then multiplied all her data by 3:

	Nearsighted	Not Nearsighted
GPA < B+	42	45
GPA ≥ B+	51	24

"I assumed that I originally picked a perfectly random sample, and that if I were to have polled 3 times as many people, my data would have been greater in magnitude, but still distributed on the same normal distribution." Does this procedure make sense to you? How do you think it affected her p-value?

12.10 One hundred students were asked this question:

> *You have a coach ticket for a flight from New York to London. Because the flight was overbooked, randomly selected passengers will be allowed to sit in open first-class seats. You are the first person selected. Would you rather sit next to: (a) Joe Biden; (b) Jill Biden; or (c) Michael Jordan?*

Here are the results:

	Joe Biden	Jill Biden	Michael Jordan
Females	12	25	10
Males	13	0	40

a. What is the most appropriate null hypothesis?
b. Are the differences between female and male responses substantial and statistically persuasive?

12.11 An online college statistics class asked students to apply a chi-square test to these observed and expected values:

Observed	41	38	20	41	27
Expected	39	37	19	38	26

Why is this question flawed?

12.12 Seventy-two randomly selected college students were asked to taste three unlabeled chocolate chip cookies and identify the one they liked best: 13 students chose Chips Ahoy; 12 chose Lady Lee, and 47 chose supermarket bakery cookies. Are these differences substantial and statistically persuasive? What is your null hypothesis?

12.13 On a full rack of billiard balls, 99 break shots resulted in 54 sunk balls, allocated among the 15 balls as follows:

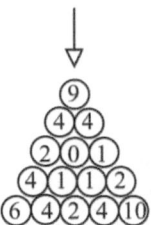

Dividing the rack into three categories (corner, adjacent, and other), is there a substantial and statistically persuasive relationship between a ball's position in the rack and its likelihood of being sunk on the break?

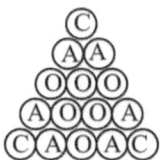

12.14 A Russian statistician named Ladislaus Bortkiewicz investigated whether the number of soldiers killed by horse kicks was randomly distributed among Prussian cavalry corps. He collected data for 122 deaths over a 20-year period in 12 cavalry corps. Using the Poisson distribution, he calculated the expected number of cases in which a corps would have no deaths in a single year (108.67), the expected number of cases in which a corps would have a single death (66.29), and so on. He compared these expected numbers to the actual number of cases. It has been suggested that the correspondence between the predicted and observed instances is too good to be true. How would you test this?

Number of Deaths by Kicks	Predicted Instances	Observed Instances
0	108.67	109
1	66.29	65
2	20.22	22
3	4.11	3
4	.63	1
5	.08	0
6	.01	0

12.15 A study of first drafts of student papers tabulated how frequently letters were used and the frequency with which typographical errors occurred. For example, "Firday" is an i typo and an r typo; "Froday" or "Frday" is an i typo; and "Friuday" is a u typo. The data were grouped according to the row on the keyboard:

	Used	Errors
Top row of letters	14,961	27
Second row of letters	9,324	21
Third row of letters	4,175	16
Space bar	5,881	11
	34,341	75

Calculate the expected number of errors for each of these four categories, assuming that the 75 errors are distributed in accord with the frequency with which letters are used. Are the differences between the expected and actual number of errors substantial and statistically persuasive?

12.16 A researcher suspected that a disproportionate number of people might be born in some months (like November, which is 9 months after Valentine's Day). Use the following data on the birth months of

148 college students to test at the 5% level the null hypothesis that all birth months are equally likely:

January	7
February	15
March	16
April	6
May	7
June	12
July	8
August	16
September	11
October	12
November	21
December	17
Total	148

12.17 A university study examined the relationship between a woman's shoe size and whether the woman frequently experienced foot pain. Are the observed differences substantial and statistically persuasive?

Shoe Size	Number Surveyed	Number with Foot Pain
$X < 8$	150	87
$8 \leq X < 10$	130	104
$X \geq 10$	120	101

12.18 The Mandarin, Cantonese, and Japanese pronunciation of "four" and "death" are almost identical, and many Chinese and Japanese people consider the number 4 to be unlucky. Californians have some say in the last four digits of their telephone numbers, and a study of the last four digits of the telephone numbers of 1984 Chinese and Japanese California restaurants counted the number of times the digits 0–9 appeared. Are the differences among the digits substantial and statistically persuasive?

Digit	0	1	2	3	4	5	6	7	8	9
Phones	960	834	789	725	562	650	726	657	1332	701

12.19 It has been argued that people may be able to postpone their deaths until after the celebration of an important event. Here are data for Jewish members of Sinai Memorial Chapel in San Francisco who died within 4 weeks of Passover between January 1, 1987, and December 31, 1995. Do these data provide substantial and statistically persuasive support for the theory?

				Weeks before (–) or after (+) Passover				
	–4	–3	–2	–1	1	2	3	4
Deaths	119	102	90	99	100	97	78	88

12.20 A student observer attended eight different college classes and recorded the number of female and male students in each classroom, and the sex of each student who made a spoken remark, question, or comment directed at the professor. She found that there were 88 females and 107 males in these classes and a total of 276 student exchanges with the professor, of which 138 were made by females and 138 by males. Are these observed differences substantial and statistically persuasive?

12.21 A survey conducted by the National Opinion Research Center asked 756 randomly selected Americans if too much or too little money was being spent on "welfare," while a random sample of 761 Americans were asked if too much or too little money was being spent on "assistance to the poor." Are the differences in the results substantial and statistically persuasive?

	"Welfare"	"Assistance to Poor"
Too little	166	484
About right	261	167
Too much	280	69
Don't know	49	41

12.22 In order to see whether women are more successful at single-sex or coeducational colleges, samples of women attending a women's college and a coeducational college were asked, "Do you feel you are successful at your college?" The results were as follows:

	Women's College	Coeducational College
Yes	37	30
No	15	13

The researcher explained that "A chi-square test was used to see if the results are statistically significant. We assume that a positive answer is a valid measure of true success, so that the null hypothesis is that the probability equals 0.5, because there is a 50-percent chance of agreeing that either yes, one is successful or no, one is not successful." The observed values were compared to these expected values:

	Women's College	Coeducational College
Yes	26	21.5
No	26	21.5

Explain why this procedure is not persuasive, and then make an appropriate statistical test.

12.23 Eighty-five randomly selected college students were asked if they had a serious romantic relationship in the past 2 years and, if so, to identify the month in which the most recent relationship began.

	Jan	Feb	Mar	Apr	May	Jun
Observed	10	13	7	3	4	4

	Jul	Aug	Sep	Oct	Nov	Dec
Observed	6	4	6	11	8	9

A chi-square test was used to test the null hypothesis that each month is equally likely for the beginning of a romantic relationship. Identify three distinct errors in the conclusion (do not check the math):

> *The expected value for each month is 85/12 = 7.083. The chi-square value is 15.66 and the two-sided p-value is .154. Therefore, the deviations between the observed and expected values are too small to be explained plausibly by random sampling error. The results are statistically substantial because there are more than 5 degrees of freedom.*

12.24 A professor showed 56 students four playing cards, numbered 1, 2, 3, and 4; shuffled them behind his back; and then selected one of the cards. While he looked at the selected card, the students tried to read his mind and wrote down their best guess as to which card was selected. Here are the results:

Card number	1	2	3	4
Number of students	10	16	27	3

What is wrong with this calculation of the chi-square statistic and why does it matter?

$$\chi^2 = \frac{\left(\frac{10}{56} - \frac{1}{4}\right)^2}{1/4} + \frac{\left(\frac{16}{56} - \frac{1}{4}\right)^2}{1/4} + \frac{\left(\frac{27}{56} - \frac{1}{4}\right)^2}{1/4} + \frac{\left(\frac{3}{56} - \frac{1}{4}\right)^2}{1/4}$$
$$= .3954$$

12.25 A researcher looked at male and female choices between Coke and Diet Coke and obtained a chi-square value of .0383: "This means that I can reject the null hypothesis that men and women are equally likely to prefer Coke or Diet Coke, since 0.0383 is less than 0.05." What is incorrect about this conclusion?

12.26 Two professors looked at the birth months of 76 major league baseball players who committed suicide. They calculated the adjusted number of suicides in each birth month by dividing the number of suicides in that month by the total number of players with that birth month and multiplying by 1,000. For example, there were a total of 638 players with January birth months; so, the adjusted January number is 1,000 (6/638) = 9.4, rounded off to 9. The expected value in each month was 126/12 = 10.5 and their calculated chi-square value was

$$\chi^2 = \frac{(9-10.5)^2}{10.5} + \frac{(13-10.5)^2}{10.5} + ... + \frac{(10-10.5)^2}{10.5} = 43.1$$

	Actual	Adjusted	Expected
January	6	9	10.5
February	7	13	10.5
March	5	8	10.5
April	5	10	10.5
May	5	9	10.5
June	6	11	10.5
July	2	3	10.5
August	19	29	10.5
September	5	8	10.5
October	7	11	10.5
November	3	5	10.5
December	6	10	10.5
Total	76	126	126

a. What is most serious error with their statistical procedure?
b. Do you think that their mistake increased or decreased the chi-square value?
c. How would you calculate the chi-square value?

12.27 Explain why the proposed test is wrong and identify the correct test:

> *I used a chi-square test to look at the relationship between high school graduation rates in California counties and the per capita income in the past 12 months in these counties. Here are my results for the first four counties alphabetically in California:*

	Alameda	Alpine	Amador	Butte
Graduation rate	86	91	88	87
Per capita income ($)	35,434	27,135	26,969	23,556

> *Graduation rate is the percent of persons age 25+ who graduated from high school and per capita income is annual income (in 2012 dollars) for the years 2008 through 2012. The chi-square value is 8.58 and the p-value is .0353, which means we would reject the null at the 5% level, but not at the 1% level.*

12.28 Two researchers collected birthday and deathday data for 120 randomly selected deceased celebrities. These data were divided into three categories: deaths that occurred during the 30 days preceding the birthday; deaths that occurred on the birthday or during the 29 days following the birthday; and deaths on other days. A chi-square statistic was used to test the null hypothesis that a person's deathday is not related to his or her birthday:

Deathdate	Number Observed	Number Expected
During 30 days preceding birthday	13	40
On or during 29 days after birthday	8	40
Other	99	40
Total	120	120

$$\chi^2 = \frac{(13-40)^2}{40} + \frac{(8-40)^2}{40} + \frac{(99-40)^2}{40} = 130.85$$

What fundamental problem do you see with this analysis?

12.29 A researcher surveyed 104 of the 1,579 students at a college to see if religious beliefs varied by grade level, suspecting that students who attend college become less religious between their freshman and senior years:

	Believing	Agnostic	Unbelieving	Total
Freshmen	11	7	12	30
Sophomore	8	9	8	25
Junior	17	5	6	28
Senior	10	6	5	21
Total	46	27	31	104

He did not find a statistically significant relationship, so he "used the sample mean to estimate the population mean;" for example, (1,579/104) 11 = 167. His chi-square value was now 102.29, which is statistically significant at the 5% level. What is the flaw in his procedure?

	Believing	Agnostic	Unbelieving	Total
Freshmen	167	106	182	455
Sophomore	121.5	137	121.5	380
Junior	258	76	91	425
Senior	152	91	76	319
Total	698.5	410	470.5	1579

12.30 For several years, the manufacturer of plain M&M's claimed that the overall percentages of the candies of different colors were as follows: 30% brown, 20% red, 20% yellow, 10% blue, 10% green, and 10% orange. A randomly selected 17.6 ounce bag of plain M&M's yielded the following data: 34% brown, 18% red, 21% yellow, 12% blue, 7% green, and 8% orange. The researchers used this chi-square statistic:

$$\chi^2 = \frac{(34-30)^2}{30} + \frac{(18-20)^2}{20} + \ldots + \frac{(8-10)^2}{10} = 2.48.$$

The p-value was .78. Their conclusion: "The probability of every color matching the null hypothesis was fairly high, and could not be rejected at the 5% level."

a. What is the correct null hypothesis?

b. Identify the error in their calculation of the chi-square statistic. Explain logically why their chi-square statistic is wrong. Do not just say, "They should have used this formula." Explain why the correct formula is more sensible than the formula they used.

Answers

12.1 The natural null hypothesis is that each of the four numbers has an equal probability of being selected, in which case the expected value for the number of students is 10 for each card. The chi-square value is

$$\chi^2 = \frac{(2-10)^2}{10} + \frac{(13-10)^2}{10} + \frac{(21-10)^2}{10} + \frac{(4-10)^2}{10} = 23.00.$$

With $4 - 1 = 3$ degrees of freedom, the p-value is minuscule. The differences between the observed and expected values seem substantial.

12.2 A chi-square test is appropriate here. Under the natural null hypothesis that each play is equally likely, the expected number in each category is 40. The chi-square value is:

$$\chi^2 = \frac{(27-40)^2}{40} + \frac{(54-40)^2}{40} + \frac{(39-40)^2}{40} = 9.15.$$

With $3 - 1 = 2$ degrees of freedom, the p-value is .0102. The observed differences seem substantial, particularly the play of scissors twice as often as rock.

12.3 Putting the data into a contingency table:

	Adult Females	Adult Males	Children	Total
Saved	296	146	56	498
Not saved	106	659	52	817
Total	402	805	108	1,315

Here are the expected values, assuming independence:

	Adult Females	Adult Males	Children	Total
Saved	152.24	304.86	40.9	498
Not saved	249.76	500.14	67.1	817
Total	402	805	108	1,315

There were far more women, somewhat more children, and far fewer men saved than would be true with independence. (Overall, 74% of the adult women, 18% of the adult men, and 52% of the children were saved.) The chi-square value is 360.7:

$$\chi^2 = \frac{(296 - 152.24)^2}{152.24} + \dots + \frac{(52 - 67.1)^2}{67.1} = 360.7.$$

The probability of such a large chi-square value with $(3 - 1)(2 - 1) = 2$ degrees of freedom is virtually zero.

12.4 We can multiply the true probabilities by 24 to determine the expected value of the number of students whose longest streak would be in each of these three categories. The chi-square distribution can then be used for a statistical test:

$$\chi^2 = \frac{(4 - .174(24))^2}{.174(24)} + \frac{(17 - .361(24))^2}{.361(24)} + \frac{(3 - .465(24))^2}{.465(24)} = 13.99.$$

With $3 - 1 = 2$ degrees of freedom, the p-value is .0008. There were far fewer than expected students guessing streaks longer than 3. One explanation is they believe in the fallacious law of averages.

12.5 The observed and expected values are

	Observed	Expected
Pitcher	51	60
Catcher	116	60
Infielder	262	240
Outfielder	111	180

Catchers are greatly overrepresented; outfielders are greatly underrepresented. The chi-square value is 82.08, and, with $4 - 1 = 3$ degrees of freedom, the p-value is less than 1.0×10^{-10}.

$$\chi^2 = \frac{(51 - 60)^2}{60} + \frac{(116 - 60)^2}{60} + \frac{(262 - 240)^2}{240} + \frac{(111 - 180)^2}{180} = 82.08$$

We could instead base the expected values on the number of players on rosters instead of the number of playing positions. If we did, pitchers would look much worse because teams carry a lot of pitchers on their rosters.

12.6 The expected value for each position is .2 (167) = 33.4. There are substantially more point guards and fewer centers than expected under the null hypothesis. The chi-square value is 38.84 and, with 5 − 1 = 4 degrees of freedom, the p-value is less than .000000001.

$$\chi^2 = \frac{(61-33.4)^2}{33.4} + \frac{(42-33.4)^2}{33.4} + \ldots + \frac{(18-33.4)^2}{33.4} = 38.84.$$

12.7 The expected values are

Forward	(3/11) 70 = 19.09
Midfielder	(3/11) 70 = 19.09
Defense	(4/11) 70 = 25.45
Goalie	(1/11) 70 = 6.36

There are nearly twice as many midfielders as expected and fewer managers than expected from the other positions. The chi-square value is 23.87, and with 4 − 1 = 3 degrees of freedom, the p-value is less than 1.0×10^{-10}.

$$\chi^2 = \frac{(14-19.09)^2}{19.09} + \frac{(36-19.09)^2}{19.09} + \ldots + \frac{(0-6.36)^2}{6.36} = 23.87.$$

12.8 Using a chi-square test with a 2 × 2 table:

	2 Choices	3 Choices	Total
Online	68	16	84
Print/Online	32	84	116
Total	100	100	200

The expected values are:

	2 Choices	3 Choices	Total
Online	42	42	84
Print/Online	58	58	116
Total	100	100	200

The chi-square value is 55.5 and the p-value is less than .000000001.

$$\chi^2 \frac{(68-42)^2}{42} + \frac{(16-42)^2}{42} + \frac{(32-58)^2}{240} + \frac{(84-58)^2}{58} = 55.5.$$

Alternatively, a difference in proportions test gives a Z-value of 7.4499. This Z-value squared is 55.5 (the chi-square value) and the two-sided p-value is equal to the p-value for the chi-square test.

12.9 a. The expected values under the null hypothesis that nearsightedness and GPA are independent are:

	Nearsighted	Not Nearsighted	Total
GPA < B+	16.65	12.35	29
GPA > B+	14.35	10.65	5
Total	31	23	54

The chi-square is 2.15:

$$\chi^2 = \frac{(42-16.65)^2}{16,65} + \frac{(45-12.35)^2}{12.35} + \ldots + \frac{(24-10.65)^2}{10.65} = 2.14.$$

With $(2-1)(2-1) = 1$ degree of freedom, the p-value is .114.

b. Tripling her data will triple each observed and expected value and triple her chi-square value to 6.41 because the numerators are squared and the denominators are not, and will reduce her p-value (here, to .0113). A chi-square test doesn't assume that the data come from a normal distribution and her procedure is not logical. We shouldn't invent data by assuming new data would be identical to the original data. That's like saying that I flipped a coin and got a head; so, if I flip the coin 10 more times, I will get 10 heads.

12.10 a. The natural null hypothesis is that sex and choice are independent.

b. A chi-square test is appropriate. We use the row and column totals

	Joe Biden	Jill Biden	Michael Jordan	Total
Females	12	25	10	47
Males	13	0	40	53
Total	25	25	50	100

to determine the expected values:

	Joe Biden	Jill Biden	Michael Jordan	Total
Females	25(47/100)	25(47/100)	50(47/100)	47
Males	25(53/100)	25(53/100)	50(53/100)	53
Total	25	25	50	100

or

	Joe Biden	Jill Biden	Michael Jordan	Total
Females	11.75	11.75	23.50	47
Males	13.25	13.25	26.50	53
Total	25	25	50	100

The chi-square value is 42.83 and the p-value is minuscule:

$$\chi^2 = \frac{(12-11.75)^2}{11.75} + \frac{(25-11.75)^2}{11.75} + \ldots + \frac{(40-26.50)^2}{26.50} = 42.83.$$

There are substantial differences between female and male preferences for Jill Biden and Michael Jordan.

12.11 Since every observed value is larger than the corresponding expected value, the sum of the observed values is not equal to the sum of the expected values. They should be equal.

12.12 The natural null hypothesis is that each of these three brands is equally likely to be picked by a randomly selected student. If so, the expected values are 72/3 = 24 for each brand. The chi-square value is 36.03, and with 3 – 1 = 2 degrees of freedom, the probability of such a large (or larger) chi-square value is less than .0000000001. There is a strong preference for supermarket bakery cookies.

$$\chi^2 = \frac{(13-24)^2}{24} + \frac{(12-24)^2}{24} + \frac{(47-24)^2}{24} = 36.03.$$

12.13 For the three categories, 3/15 of the balls are in the corners, 6/15 are adjacent, and 6/15 are other, implying these expected values:

	Observed	Expected
Corner	25	(3/15) (54) = 10.8
Adjacent	22	(6/15) (54) = 21.6
Other	7	(6/15) (54) = 21.6
Total	54	54

Far more corner balls were sunk and far fewer other balls were sunk than would be expected if each ball were equally like to be sunk. The chi-square value is 28.55 and, with 3 – 1 = 2 degrees of freedom, the p-value is .0000007.

$$\chi^2 = \frac{(25-10.8)^2}{10.8} + \frac{(22-21.6)^2}{21.6} + \frac{(7-21.6)^2}{21.6} = .022.$$

12.14 We can use a chi-square test. The chi-square value is .7889. With 7 – 1 = 6 degrees of freedom, the p-value is .9924, which is suspiciously high:

$$\chi^2 = \frac{(109-108.67)^2}{108.67} + \frac{(65-66.29)^2}{66.29} + \ldots + \frac{(0-.01)^2}{.01} = .7889.$$

12.15 The expected values are determined by multiplying the frequencies by 75; for example, for the top row, 14,961/34,341) 75 = 32.67. Here are the results:

	Expected Errors	Actual Errors
Top row of letters	32.67	27
Second row of letters	20.36	21
Third row of letters	9.12	16
Space bar	12.84	11
Total	75	75

The most striking difference is for the third row of letters. We can use a chi-square test to compare the observed and expected values:

$$\chi^2 = \frac{(27-32.67)^2}{32.67} + \frac{(21-20.36)^2}{20.36} + \frac{(16-9.12)^2}{9.12} + \frac{(11-12.84)^2}{12.84} = 6.44.$$

With $4-1=3$ degrees of freedom, the p-value is .092.

12.16 The expected value for each month is $148/12 = 12.333$, and the chi-square value is 20.162. With $12-1=11$ degrees of freedom, the p-value is .0431:

$$\chi^2 = \frac{(7-12.33)^2}{12.33} + \frac{(15-12.33)^2}{12.33} + \ldots + \frac{(17-12.33)^2}{12.33} = 20.162.$$

12.17 Overall, 292 of 400 (73%) experienced foot pain. If there were no relationship between shoe size and foot pain, we expect 73% of those in each size category to experience foot pain:

Shoe Size	Surveyed	Observed	Expected
$X < 8$	150	87	.73(150) = 109.5
$8 \leq X < 10$	130	104	.73(135) = 94.9
$X \geq 10$	120	101	.73(120) = 87.6
	400	292	292

The chi-square value with $3-1=2$ degrees of freedom is 7.546 and the p-value is .0228.

$$\chi^2 = \frac{(87-109.5)^2}{109.5} + \frac{(104-94.9)^2}{94.9} + \frac{(101-87.6)^2}{87.6} = 7.546.$$

By Fisher's 5% rule, the observed differences are statistically significant. The differences between the smallest and largest shoe sizes seem substantial to me.

12.18 There are far fewer 4s and far more 8s (a lucky number in Chinese and Japanese cultures) than other digits. There are a total of 7,936 observations and, under the null hypothesis, the expected value for each of the ten digits is $7{,}936/10 = 793.6$. The chi-square statistic is

$$\chi^2 = \frac{(960-793.6)^2}{793.6} + \frac{(834-793.6)^2}{793.6} + \ldots + \frac{(701-793.6)^2}{793.6} = 541.82.$$

With $10 - 1 = 9$ degrees of freedom, the p-value is less than .000000001.

12.19 We can use a chi-square test. There were a total of 773 deaths. If a death during this 8-week period is equally likely to be in any week, the expected value for each week is $773/8 = 96.625$. The chi-square value is

$$\chi^2 = \frac{(119-96.625)^2}{96.625} + \ldots + \frac{(88-96.625)^2}{96.625} = 10.472.$$

With $8 - 1 = 7$ degrees of freedom, the p-value is .1633, not sufficient to reject the null hypothesis at the 5% level. In addition, there were more deaths in the 4 weeks before Passover than during the 4 weeks after Passover, the opposite of what was predicted.

12.20 We can use a chi-square test:

	Observed	Expected
Females	138	(88/195) 276 = 124.55
Males	138	(107/195) (276) = 151.45
Total	276	276

Females made more comments than expected though it is debatable whether the sex differences are substantial

The chi-square value is 2.647; with 1 degrees of freedom, the p-value is a non-persuasive .1037.

$$\chi^2 = \frac{(138-124.55)^2}{124.55} + \frac{(138-151.45)^2}{151.45} = 2.647.$$

12.21 Here are the expected values:

	"Welfare"	"Assistance to Poor"	Total
Too little	323.93	326.07	650
About right	213.30	214.71	428
Too much	173.93	175.08	349
Don't know	44.85	45.15	90
Total	756	761	1,517

Respondents were far more likely to say "too little" and far less likely to say "too much" when the question said "assistance to the poor" instead of "welfare." With $(4 - 1)(2 - 1) = 3$ degrees of freedom, the chi-square value is 304.49 and the p-value is minuscule.

$$\chi^2 = \frac{(166 - 323.93)^2}{323.93} + \frac{(41 - 45.15)^2}{45.15} = 304.49.$$

12.22 There is no good reason why yes and no answers should be equally likely. Plus, this test doesn't get at the central question, whether women who are at women's colleges feel more successful than those at coeducational colleges. A more plausible test is whether the yes/no answers are independent of the college attended. The expected values, assuming independence, are as follows:

	Women's College	Coed College	Total
Yes	36.67	30.32	67
No	15.32	12.67	28
Total	52	43	95

For example, $(52)(67)/105 = 36.67$. The observed values are almost exactly equal to the expected values. The chi-square value is virtually zero and the p-value with $(2 - 1)(2 - 1) = 1$ degree of freedom is .88:

$$\chi^2 = \frac{(37 - 36.67)^2}{36.67} + \frac{(15 - 15.32)^2}{15.32} + \ldots + \frac{(13 - 12.67)^2}{12.67} = .022.$$

12.23 The chi-square test is one-sided and the statement should say that the deviations between the observed and expected values are small enough to be explained plausibly by random sampling error. Whether the observed differences are substantial would be gauged by subjectively judging the difference between the observed and expected values.

12.24 This calculation does not take the sample size into account. The percentage differences are more statistically persuasive, the larger the sample. Here, the chi-square value is 56 times larger: 56 (.3954) = 22.14:

$$\chi^2 = \frac{(10 - 14)^2}{14} + \frac{(16 - 14)^2}{14} + \frac{(27 - 14)^2}{14} + \frac{(3 - 14)^2}{14}$$

$$= 56(.3954) = 22.14.$$

12.25 The value .0383 is the chi-square value, not the p-value.

12.26 Their actual numbers are incorrect, but that is not the issue here.

　a. They should use the total number of suicides to calculate the expected values, not to adjust the observed values.

　b. By increasing the total number of observed values from 76 to 126, they increased the chi-square value.

c. I would leave the original 76 observed values as is and use the total number of births in each month to calculate the expected values; for example, for January, the expected value is 76 (638/ total number of players).

12.27 The numbers in each cell of a contingency table should be the number of counties in that category, not the values of the graduation rate and income. The chi-square should be something like this

	Low Income	Middle Income	High Income
Low graduation	5	7	1
High graduation	0	8	5

It would be better to do a regression model.

12.28 Each of the first two categories contain 30 days, but the third category contains 305 days (not counting leap year). If deathday is unrelated to birthday, every day is equally likely and we expect far more deaths in a category with 305 days than in a category with only 30 days.

12.29 We cannot assume that a larger sample will be identical to the current sample, only bigger. This is analogous to flipping a coin once, obtaining a tail, and assuming that 15 flips will yield 15 tails.

12.30 a. The null hypothesis is that the contents of a bag of plain M&M's constitute a random sample from a population with the candy colors distributed according to the manufacturer's claims. Thus a randomly selected M&M has a 30% chance of being brown, a 20% chance of being red, and so on.

b. They should have used the number of observed and expected M&M's, not the percentages. Logically, their statistic is flawed because it does not take into account the number of M&M's in the sample. For example, these could be 34% and 30% of 100 M&M's or 100,000 M&M's. An observed 4-percentage-point difference is more persuasive evidence against the null hypothesis when the sample size is large.

13

Simple Regression

13.1 This regression equation was estimated using data on college applicants who were admitted both to Pomona College and to another college ranked among the top 20 small liberal arts colleges by *US News & World Report*:

$$Y = .2935 + .0293X, \ R^2 = .562$$
$$(.0781) \quad (.0065)$$

where:

Y = fraction of students who were admitted to both this college and Pomona College who enrolled at Pomona college; average value = .602

X = *US. News & World* Report ranking of this college

( ) = standard error

a. How were the numbers .2935 and .0293 obtained? (Do not show any formulas; explain in words the basis for the formulas.)

b. Explain why you are not surprised that the R-squared for this equation is not 1.0.

c. Does the coefficient of X have a plausible value?

d. Is the estimated coefficient of X statistically persuasive?

e. What is the null hypothesis in (d)?

f. What is the predicted value of Y for X = 30?

g. Why should we not take the prediction in (f) seriously?

13.2 Explain why you either agree or disagree with each of these claims about the simple regression model:

a. If the estimated slope in a simple regression equation is negative, it cannot be statistically significant.

b. The estimated correlation coefficient doesn't depend on which variable is the dependent variable and which is the explanatory variable.

c. Ordinary least squares assumes that the explanatory variable is normally distributed.

DOI: 10.1201/9781003630159-13

13.3 Two researchers used annual data for 2000–2021 to estimate a simple regression model, where X = percentage change in the S&P 500 from January 1 through January 31 and Y = percentage change in the S&P 500 index of stock prices from February 1 through December 31 of that same year:

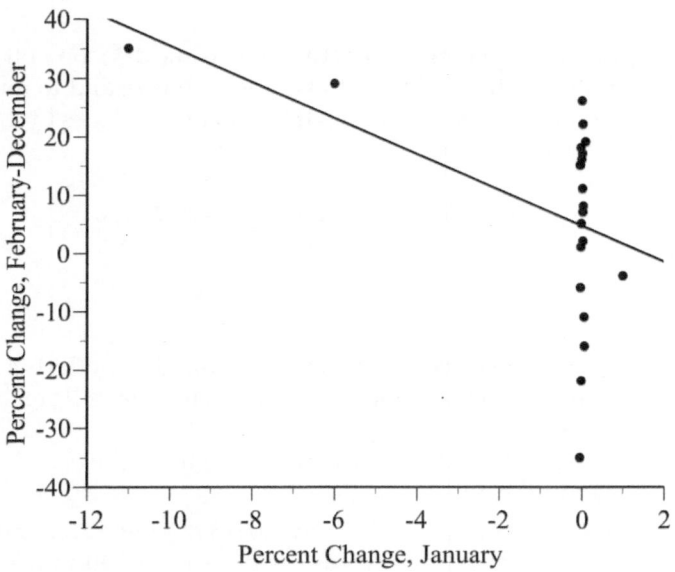

The authors reported that

> $Y = 4.670 - 3.081X$, $R^2 = .22$, $b = -3.08$, *t-value* $= -2.39$, *p-value* $= .1917$. The beta is not substantial because the t-value is less than 2. We can reject the null hypothesis. Our results show that February-December % change in the S&P 500 does not depend on the % change in January.

Identify five distinct problems with this analysis.

13.4 The equation $Y = \alpha + \beta X + \varepsilon$ was estimated by ordinary least squares. Explain why you either agree or disagree with these statements:

a. Least squares regression minimizes the sum $\sum_{i=1}^{n}(X_1 - \bar{X})^2$

b. R^2 is equal to the correlation coefficient squared.

c. A doubling of the number of observations, with the new data exactly replicating the original data, would increase R^2.

d. $R^2 = 1$ means $Y = X$.

13.5 A researcher wanted to investigate burnout in competitive mountain biking. He looked at every California U23 (ages 19–22) mountain biker in 2012, and calculated the total number of sanctioned events they had competed in while in U23 and while they were in Junior racing (ages 17–18). The least squares regression was Y = 1.45 + 1.11X, with t = 7.91 for the slope estimate and R-squared = .5.

  a. What are the most important problem(s) with these data?

  b. Which variable in the regression equation should be the dependent variable?

  c. Is the two-sided p-value for the slope estimated less than .05?

  d. Explain why you either agree or disagree with this interpretation of the regression results: "This shows that racers generally raced close to the same number of times as a junior and as a U23."

13.6 A researcher specified this time-series model

$$Y = \alpha + \beta X + \varepsilon$$

where:
  Y = annual percentage change in household spending
  X = annual percentage change in the S&P 500 index of stock prices

He reported these results

$$\text{Estimate of } \alpha = .21$$

$$\text{Estimate of } \beta = .48$$

$$\text{Estimate of } \varepsilon = 1.13$$

$$t = 2.43 \text{ for test of } H_0: X = 0:$$

  What two mistakes do you see in these reported results?

13.7 Data were collected for 25 new cars on the car's weight (X) and estimated highway miles per gallon (Y). The model Y = α + βX + ε was estimated by ordinary least squares in three different ways: (a) the data were arranged alphabetically by car name; (b) the data were arranged numerically from the lightest car to the heaviest; and (c) the data were arranged numerically from the lowest miles per gallon to the highest. Which procedure yielded the smallest estimate of the intercept? Which yielded the highest estimate?

13.8 The equation $Y = \alpha + \beta X + \varepsilon$ was estimated by ordinary least squares. How would a doubling of the number of observations, with the new data exactly replicating the original data, affect the

   a. estimates of the slope and intercept?

   b. standard error of the estimate of the slope?

   c. t-value of the estimate of the slope?

   d. R^2?

13.9 A regression model was constructed to predict the sale price of homes in Elm City:

> We looked at the list (asking) prices for 60 homes that are for sale in Elm City. We wrote down the square footage and price of the property and used that information to make a scatterplot with a line of least regression. The relationship between square footage and price of the listing was positive. The equation is y = 242 + .00247x, where y = square footage and x = list price.

How would you improve this study and the report of the results?

13.10 Critically evaluate:

> Many people believe that the 1979 spike in oil prices increased the unemployment rate in the United States. However, when I used monthly data for 1979 to estimate the equation U = α + βP + ε, where U is the unemployment rate and P is the price of Texas crude oil, the estimated slope was not statistically significant—demonstrating that oil prices, in fact, had no effect on the unemployment rate.

13.11 A researcher used least squares to estimate the equation $Y = \alpha + \beta X + \varepsilon$, where Y = college GPA and X = high school GPA. First, he estimated the equation using all his data. Then, he separated his data into male and female students and estimated separate equations for each sex. Identify three apparent errors you see in these reported results.

	Men	Women	Total
Sample size	407	482	869
Intercept	−.22	−.04	−.12
	(−.32)	(−.24)	(−.08)
Slope	.88	.80	.83
	(−.32)	(−.24)	(−.08)
R-squared	1.01	1.54	1.27

( ): standard error

13.12 A 2005 linear regression estimated the relationship between the dollar prices P of 45 used 2001 Audi A4 2.8L sedans with manual transmissions and the number of miles M the car had been driven:

$$P = 16,958 - .0677M , \quad R^2 = .70$$
$$(2,553) \qquad (.0233)$$

The standard errors are in parentheses.

a. Does the value 16,958 seem reasonable?

b. Does the value –.0677 seem reasonable?

c. Is the estimated relationship between M and P statistically persuasive?

d. Should the variables be reversed, with M on the left-hand side and P on the right-hand side?

e. Suppose that the true relationship is $P = \alpha - \beta_1 M + \beta_2 C + \varepsilon$, where C is the car's condition (scale of 1–5, with 5 best) and α, β_1, and β_2 are all positive. If M and C are negatively correlated, does the omission of C from the estimated equation bias the estimate of β_1 upward or downward?

13.13 Here is an excerpt from a debate between two famous economists on the relationship between the growth of the money supply and the rate of inflation. Which economist do you agree with on the question of whether regression implies causation?

Steve Hanke: I related the growth rate in the money supply to the inflation rate [in] 147 countries and the money supply measure I used is M2 and what do we end up with? We have almost a perfect correlation.

John Cochrane: Correlation is not causation. We've known that forever.

Steve Hanke: I also ran a regression and that does imply causation.

John Cochrane: Oh, no no. Correlation does not imply causation. Come on now.

13.14 A researcher obtained the Scholastic Aptitude Test (X) scores and college grade point averages (Y) of 396 college students and reported these regression results:

$$Y = 1146.0 + 1.4312X \, , \, R2 = 0$$
$$(195.6) \quad (71111.6)$$
$$[5.86] \quad [.02]$$

( ): std. devs.

[ ]: t-values

a. There is a major error in these results. What is it?

b. The researcher concluded that, "Therefore I reject the null hypothesis and conclude that SAT scores and grades are not related." What is the basis for this conclusion and why is it wrong?

13.15 A study of 15 stocks that were wildly popular in the 1970s collected these data: (1) the December 1972 price earnings ratio (P/E) and (2) the annual percentage rate of return (R) from 1973 through 2000. What statistical test would you use to see if stocks with high P/E ratios tended to have low annual returns? (You do not have to actually do the test, just identify it clearly.) Be sure to identify the null hypothesis and how you would test it.

	1972 P/E	R
Polaroid	90.7	−6.07
McDonald's	85.7	11.85
Walt Disney	81.6	10.58
Avon Products	65.4	6.30
Kresge (now Kmart)	54.3	1.20
Simplicity Pattern	53.1	−2.01
Xerox	48.8	−1.99
Eastman Kodak	48.2	2.68
Coca-Cola	47.6	14.64
IBM	37.4	8.65
Procter & Gamble	32.0	12.27
PepsiCo	29.3	16.17
General Electric	26.1	16.85
Philip Morris	25.9	18.00
Gillette	25.9	14.89

13.16 A finance professor estimated this model to see if investors could increase their returns by buying stocks with high PEG ratios:

The model is

$$R = \alpha + \beta PEG + \varepsilon$$

using these data for the 30 stocks in the Dow-Jones Industrial Average on December 31 of each year, during the period December 31, 2004, through December 31, 2021.

P = per share stock price on December 31 of that year

E = per share earnings during the preceding year

G = growth rate of earnings over the next 5 years

PEG = (P/E)/G

R = stock rate of return over next 7 months; January 1 through July 31

a. What are the two most important issues you would raise with this analysis?

b. Give at least three ways in which this presentation of the results could be improved:

Least squares estimates show a statistically significant relationship between PEG and R:

	Coefficient	Standard Error
Constant	.0198368	.0238002
PEG	.0050861**	.0017132

**: p < .05

13.17 US taxpayers can either take a standard deduction or an itemized deduction based on property taxes, mortgage interest, charitable contributions, and a few other items. The deduction of property taxes and mortgage interest is intended to encourage home ownership which is thought to boost active citizenship, stable neighborhoods, and other positive externalities. A cross-section regression equation was estimated using 2015–2019 average values for each of the 58 counties in California: $Y = \alpha + \beta X + \varepsilon$, where Y is the percentage of the eligible population who voted in the 2019 presidential primary election and X is the average value during the years 2015–2019 of the ratio of the number of homeowners divided by the total number of households in each county.

The estimated value of β was 19.35 with a standard error of 20.95. Explain why you either agree or disagree with the authors' conclusion that

> Although homeownership seems to have an economically significant and positive impact on voting percentage, statistical significance is lacking. Therefore, critics of the mortgage interest deduction have further support that home ownership does not encourage voting.

13.18 What is puzzling about this plot of the residuals from a least squares regression?

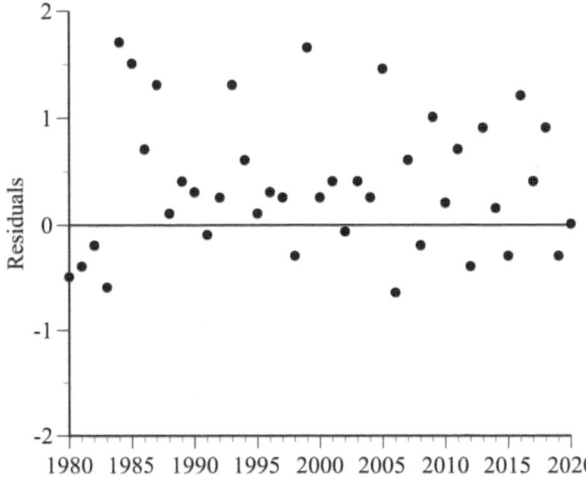

13.19 This regression equation was estimated using data for 481 California Census blocks that were more than 95% Hispanic in 2008,

$$Y = .843 + .00023X, \ R^2 = .0025$$
$$[97.12] \quad [1.18]$$

where:
   Y = percentage of the two-party vote received by Barack Obama in the 2008 Presidential election
   X = median household income of Hispanics, $1000s of dollars

The t-values are in brackets.

   Identify the logical error in this conclusion:

   Income is not statistically significant; the p-value is .27, meaning that there's a 27% chance we accept the null hypothesis that income has no effect on the Hispanic vote.

13.20 In 2005, Granite Construction applied for a permit to build a quarry near Temecula, California. Rocks would be dislodged from a mountain by daily blasts of 10,000 pounds of explosives and then crushed into gravel and sand that are used for concrete, asphalt, and other construction material. Mining and processing would go on 20 hours a day, 6 days a week, with 1,600 trucks entering and leaving the site daily.

   The proposed quarry was in a mountain gap where ocean breezes carry cool air to the Temecula Valley. Residents were alarmed about the effect of this quarry on the city's economy and property values. However, an economist hired by Granite argued that in the city of Corona, which is 40 miles away and has had mines for many years, there was a "direct positive correlation" between annual mine production and property values over the past 20 years. Temecula residents evidently should be celebrating that their city was chosen for this quarry—which would have a positive effect on property values.

   Provide a clear, persuasive refutation of this statistical argument.

13.21 A researcher used annual data for Major League Baseball (MLB) players for 1962 (when the current 162-game schedule was introduced) through 2011 to estimate this equation

$$Y = -14.75 + .083X , \ R^2 = .508$$
$$(2,285) \quad (.012)$$

where:
   Y = average number of home runs per game that season
   X = average weight (in pounds) of MLB players

and the standard errors are in parentheses.

The average value of Y is 1.76 and the average value of X is 191.76.

a. Do you agree with the author's choice of which variable should be the dependent variable?

b. Why is the estimated intercept negative?

c. Is the estimated slope substantial, plausible, and statistically persuasive?

d. What fraction of the variation in Y is explained by X?

13.22 Critically evaluate:

> The American Mustache Institute (AMI) is planning a "Million Mustache March" on the nation's capital. Their rallying cry: Pass the "Stimulus to Allow Critical Hair Expenses," or STACHE Act, which would "provide a $250 annual tax deduction for expenditures on mustache grooming supplies." According to AMI research, mustached Americans earn 4.3% more money than "clean-shaven Americans" on average per year. Therefore incentivizing mustache growth would boost the economy.
>
> "Given the clear link between the growing and maintenance of mustaches and income, it appears clear that mustache maintenance costs qualify for and should be considered as a deductible expense related to the production of income under Internal Revenue Code Section 212," wrote Dr. John Yeutter, a tax policy professor at Northeastern State University.

13.23 A researcher estimated a simple regression model using stock market returns for these (X, Y) pairs: X = 1930 return, Y = average return 1930–1939; X = 1940 return, Y = average return 1940–1949, and so on. The estimated correlation was positive, leading him to conclude that annual stock market returns could be used to predict returns over the next 10 years. What statistical flaw do you see in his procedure?

13.24 A study of the number of times that Trump tweeted the word *economy* each day and the daily high temperature in Moscow five days later found this relationship:

$$Y = 49.49 + 2.00Z , \quad R^2 = .01$$
$$[65.13] \quad [2.62]$$

where:

Y = Moscow temperature (°F)

Z = daily Z-score for Trump tweets of the word *economy*

The t-values are in brackets.

a. What is the average value of Z?

b. What is the average value of Y?

c. Does the coefficient of Z seem statistically persuasive to you?

d. Does the coefficient of Z seem substantial to you?

e. Why are you deeply skeptical of this equation?

13.25 Explain why you either agree or disagree with these statements:

a. "Two variables with a +1.00 correlation move up and down at the same time, with the same magnitude."

b. "If the estimated slope in a simple regression model is larger than the true slope, then the estimated intercept is smaller than the true intercept."

13.26 Write the correct correlation value on each graph: –.140, –.988, .509, .863:

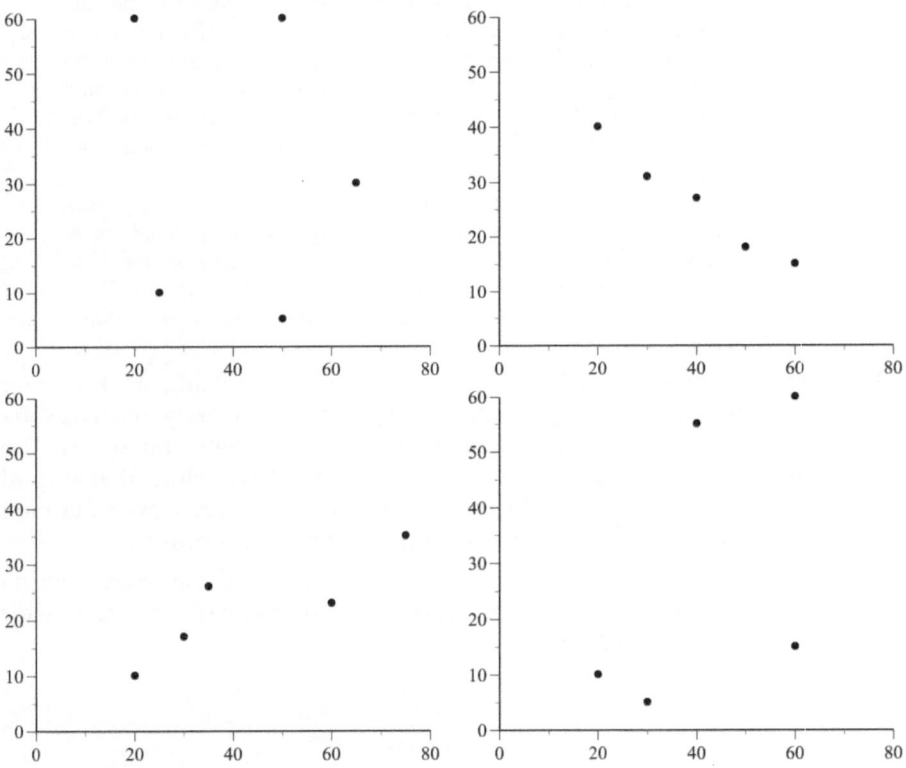

13.27 A researcher estimated this equation for 74 breakfast cereals,

$$Y = .3834 - .0074X$$
$$[16.81] \quad [3.55]$$

where:

Y = price per ounce

X = grams of sugar per ounce

The t-values are in brackets.

Explain why you either agree or disagree with this reasoning: "The Y-intercept of 0.3834 is statistically significant at the 1% level. This result makes perfect sense because one would expect a box of cereal with zero grams of sugar to cost zero cents."

13.28 A study of the effect of driving speed on traffic fatalities estimated the following model by ordinary least squares,

$$Y = \alpha + \beta X + \varepsilon,$$

where:

Y = deaths per million miles driven

X = average speed, miles per hour

The researcher used the data shown below:

Year	X	Y
1980	57.0	1.6
1984	59.2	1.4
1985	59.5	1.3

a. The researcher argued that "supporters of lower speed limits would expect β to be close to one, reflecting a strong, direct effect of X on Y." What is wrong with this argument?

b. Do you think that an ordinary least squares estimate of β using these data will be positive or negative?

c. Why, no matter what the estimate of β, are you suspicious of her data?

13.29 Least squares will be used to estimate the model $Y = \alpha + \beta X + \varepsilon$, using annual data for 1991 through 2000. If $X = 100$ and $Y = 100$ in 1991 and $X = 200$ and $Y = 200$ in 2000, can the least squares estimate of β possibly be negative? Explain your reasoning.

13.30 Least squares regression minimizes which of the following sums? Explain your reasoning.

a. $\displaystyle\sum_{i=1}^{n}(Y_i - X_i)^2$

b. $\displaystyle\sum_{i=1}^{n}(Y_i - \overline{Y})^2$

c. $\displaystyle\sum_{i=1}^{n}(X_i - \overline{X})^2$

d. $\displaystyle\sum_{i=1}^{n}(Y_i - a - bX_i)^2$

Answers

13.1 a. The intercept and slope were estimated by fitting the line that minimizes the sum of squared vertical deviations of the points from the line.

    b. There is not a perfect fit because the choice of college is affected by other factors (such as geography and special interests).

    c. The .0293 coefficient of X means that if the non-Pomona school's ranking increases by 1, say from 10 to 11 (which is a lower ranking), the chances that the student will enroll at Pomona increases by 2.93%, which seems plausible—if not particularly large.

    d. Yes, because the t-value is .0293/.0065, which is much higher than the (approximate) value of two needed for statistical significance at the 5% level.

    e. The null hypothesis is that X has no effect on Y.

    f. The predicted value of Y is .2935 + .0293(30) = 1.1725.

    g. This is incautious extrapolation, which is clearly inappropriate here since the value of Y cannot possibly be larger than 1.

13.2 a. No, the null hypothesis that the slope is zero is rejected by estimates that are several standard deviations from zero, in either direction.

    b. Yes, the formula for the correlation coefficient r is

$$r = \frac{\sum_{i=1}^{n}(x_i - \bar{x})(y_1 - \bar{y}) / (n-1)}{s_x s_y}.$$

The calculated value doesn't depend on which variable is X and which is Y.

    c. No, the least squares procedure does not assume that the explanatory variable is normally distributed.

13.3 With 22 − 2 = 20 degrees of freedom and t = −2.39, the p-value is surely not .1917. (In fact, the two-sided p-value is .027.) "Substantial" relates to the size of the beta coefficient, not the t-value. If the p-value is .1917, then the null hypothesis is not rejected at the 5% level. If the null hypothesis is rejected, this indicates that Y is related to X. The results seem to depend entirely on the two outliers. (The January changes were miscalculated. The two outliers are the correct percent changes but the other values were calculated as fractions; e.g., .04 instead of 4%.)

13.4 a. False, OLS finds the values of a and b that minimize

$$\sum_{i=1}^{n}(Y_i - a - bX_i)^2.$$

  b. True.

  c. False, no effect.

  d. False, $R^2 = 1$ means the data lie on a straight line, $Y = a + bX$, not necessarily the line $Y = X$.

13.5 a. There is survivorship bias in that these data exclude people who raced as Juniors, burned out, and did not race as U23. Also, this does not control for the fact that 22-year-olds will have had more races than 19-year-olds simply because they have been U23 longer.

  b. U23 races should be the dependent variable.

  c. Yes, because the 7.91 t-value is much larger than 2.

  d. No, the regression results show that people who raced more as Juniors tended to race more as U23. We can see from the slope and intercept that Y does not equal X.

13.6 The model doesn't estimate a single value for ε. The t-values are for the coefficients (like β), not X.

13.7 Imagine a scatter diagram. The order in which the data are plotted doesn't matter.

13.8 Imagine a scatter diagram with new points placed on top of each of the original points.

  a. The estimates of the slope and intercept do not change.

  b. The standard error of the estimate of the slope declines because the number of observations increased.

  c. The t-value of the estimate of the slope increases because the standard error decreased.

  d. The correlation is unchanged, so the R^2 is unchanged.

13.9 It would be better to use actual sale prices instead of list prices. The dependent variable should be the price, not the square footage, and there should be additional explanatory variable like the number of bathrooms and whether the home has a swimming pool. It would be helpful to show the p-values for the coefficients of the explanatory variables. I don't know what "line of least regression" means.

13.10 Not statistically significant does not prove that there is no effect. Perhaps these 12 monthly observations were not a large enough sample to obtain a low p-value, or perhaps the model was misspecified (should it be the change in oil prices?), or perhaps important variables were omitted.

13.11 The sample sizes don't add up; standard errors cannot be negative; and R-squared cannot be larger than 1.

13.12 a.   The predicted price is 16,958 if the car has never been driven. This value seems reasonable.

   b.   This equation predicts that driving an extra mile reduces the price of the car by .0677, about 7 cents/mile. This seems reasonable.

   c.   The estimated relationship between M and P is statistically significant at the 5% level because the absolute value of the t-value is larger than 2: $|t| = .0677/.0233 = 2.91$.

   d.   No, mileage affects price, not the other way around.

   e.   The absolute value of the estimate of β_1 will be biased upward. When M goes up, C tends to go down, which reduces the price. Therefore an increase in M will seem to have a larger negative effect on price.

13.13 Cochrane is right: regression does not imply causation.

13.14 a.   If X is SAT scores, the predicted values of Y are much too large to be GPAs. The data were apparently entered with Y = SAT score and X = GPA.

   b.   The null hypothesis is that SAT scores and grades are not related ($\beta = 0$). The t-value is too low (less than 2.0) to reject this null hypothesis at the 5% level, so the researcher does not reject the null hypothesis. The conclusion that "SAT scores and grades are not related" is also too strong, since a failure to reject the null hypothesis does not prove it is true.

13.15 We can use least squares regression with the rate of return as the dependent variable and the 1972 P/E as the explanatory variable, using a t-statistic to test the null hypothesis that the slope coefficient is 0.

13.16 a.   The 7-month horizon is suspicious and suggests p-hacking. The growth rate of earnings over the next 5 years is not available to investors who are making buy/sell decisions on December 31 of each year.

   b.   The estimated coefficients and standard errors should be rounded off. Instead of asterisks, the author should show the p-values. The sign and size of the coefficient of PEG should be discussed; it was expected to be negative.

13.17 The t-value, 19.35/20.95, is less than 2; however, not rejecting the null hypothesis that there is no effect does not prove that there is no effect.

13.18 The average value of the residuals should be zero but it isn't.

13.19 We can't put a probability on the null hypothesis being true unless we do a Bayesian analysis. The p-value relates to the probability of observing certain data if the null hypothesis is true, not the probability that the null hypothesis is true if we observe certain data.

13.20 When two things grow over time, there can be a statistical correlation without any causal relationship (e.g., beer sales and the number of marriages in the United States).

13.21 a. Yes, home runs may depend on weight, but not the other way around.

   b. The estimated intercept is negative because this helps the least squares line fit the data better.

   c. Yes, the estimated slope is substantial, plausible, and statistically persuasive. If X increases by 1 (about .5% at the mean), the predicted value of home runs per game increases by .083 (about 5%).

   d. Since $R^2 = .508$, 50.8% of the variation in Y is explained by X.

13.22 There is likely to be reverse causation here.

13.23 The first year is part of the 10-year period, which makes it more likely that there is a positive statistical relationship. He should have compared 1930 with 1931–1940, and so on.

13.24 a. The average value of a standardized Z-value is 0.

   b. The average value of Y is 49.49. Y is equal to its average value when X is equal to its average value, here 0.

   c. Yes, $t = 2.62$ is well above 2.

   d. A one-standard deviation increase in tweets is predicted to increase the high temperature by 2°F.

   e. This is clearly data mining, as there is no reason for a causal relationship between these two variables.

13.25 a. No, there is a perfect positive linear relation, but the slope need not be one.

   b. No, if the error terms are mostly positive, the estimated slope and intercept might both be above their true values.

13.26 Starting at the top left and going clockwise: −.140, −.988, .863, .509.

13.27 First, $p < .001$ strongly rejects the null hypothesis that the intercept is zero. Second, it doesn't make "perfect sense" that a box of sugar-less cereal would be free.

13.28 a. Supporters of lower speed limits expect β to be positive (perhaps with R^2 that is close to one). There is no persuasive reason why β should be close to one.

   b. The data show Y declines as X increases. The ordinary least squares estimate of β is negative.

   c. The peculiar choice of years 1980, 1984, and 1985 is very suspicious.

13.29 Yes; for example:

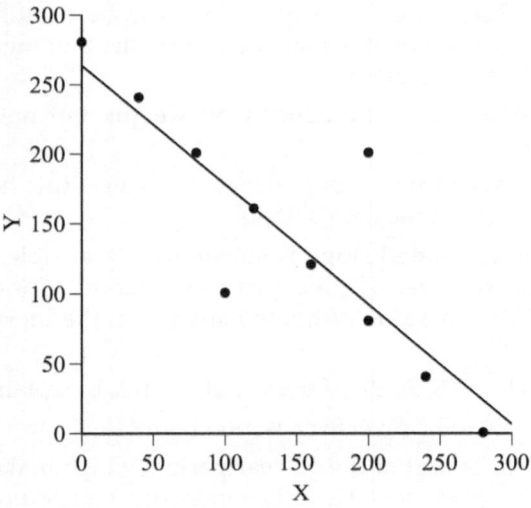

13.30 Sum (d) minimizes the sum of squared prediction errors.

14

Regression toward the Mean

14.1 Police in the United Kingdom installed speed cameras at locations where there had been a jump in traffic accidents. This policy has been justified by the fact that, after the installation of speed cameras at these dangerous locations, there was usually a decline in traffic accidents. Explain why the benefits are most likely overstated.

14.2 While working as a consultant for the Israeli Air Force in the late 1960s, Daniel Kahnemann was told by the flight instructors that students improved after being yelled at following poor performances. How do you think Kahnemann explained this contradiction of his strong belief that instructors should offer positive encouragement?

14.3 A psychology experiment had participants pretend they were teachers dealing with students who sometimes showed up late for class. Almost all the participants chose to punish those who were late and were pleased that tardy students were more punctual after being punished. However, each student's arrival time had been randomly set before the experiment began. How do you explain the results?

14.4 Explain why you either agree or disagree with this explanation of regression toward the mean by a trust investment officer:

> Key financial ratios of companies tend, over time, to revert to the mean for the market as a whole. High returns invite new entrants, driving down profitability, while poor returns cause the exit of competitors, leaving a more profitable industry for the survivors.

14.5 Explain why you either agree or disagree with Peter Bernstein, *Against the Gods*: "Regression to the mean … is what motivates the gambler's dream that a long string of losses is bound to give way to a long string of winnings."

14.6 Peter Bernstein, author of *Against the Gods*, gives this example of regression toward the mean: "Joseph had this preordained sequence of events in mind when he predicted that seven years of famine would follow seven years of plenty." Is this a good example of regression toward the mean?

14.7 Since 1937, the Gallup polling organization has been asking Americans, "Do you approve or disapprove of the way XYZ is handling his job as president?" Of the people elected president since 1937, all have had lower favorability ratings at the end of their first term than at the beginning. Provide a statistical explanation.

14.8 Sir Francis Galton observed that the adult children of unusually tall parents tended to be somewhat shorter than their parents while the

DOI: 10.1201/9781003630159-14

reverse was true of the children of unusually short parents. Does this imply that we will all soon be the same height?

14.9   Suppose that you meet a 25-year-old man who is 6 feet, 2 inches tall and he tells you that he has two brothers, one 23 years old and the other 27 years old. If you had to guess whether he is the shortest, tallest, or in-between of these three brothers, which would you select? Explain your reasoning.

14.10 *Sports Illustrated*'s cover on November 18, 1957 had a picture of the Oklahoma football team, which had not lost in 47 games, with the caption "Why Oklahoma is Unbeatable." The Saturday after this issue appeared, Oklahoma lost to Notre Dame 7–0, starting the legend of the Sports Illustrated cover jinx—the performance of an individual or team usually declines after they are pictured on the cover of *Sports Illustrated*. What statistical argument could explain the Sports Illustrated cover jinx?

14.11 Give a statistical explanation for the fact that (as of May 2023), a total of 230 golfers have won at least one of the four major men's golf championships, but 143 (62%) of these golfers only won once.

14.12 Cy Young Awards are given each year to the best pitcher in Major League Baseball's American League and National League. LaMarr Hoyt won a Cy Young Award in 1983 and complained in 1984 that he had been jinxed: "I'll tell you, there have been a lot of times this season I've felt jinxed. A lot happened to me that is unexplainable." An article reporting his grievance noted that the four previous Cy Young winners all had disappointing seasons the year after winning the award.

A modern tool for assessing baseball players is wins above replacement (WAR)—the higher, the better. The figure below shows the average WAR for all Cy Young winners 2000–2023 during their award-winning season and during the seasons before and after they won the award. Are these data consistent with the jinx theory? How would you explain these data?

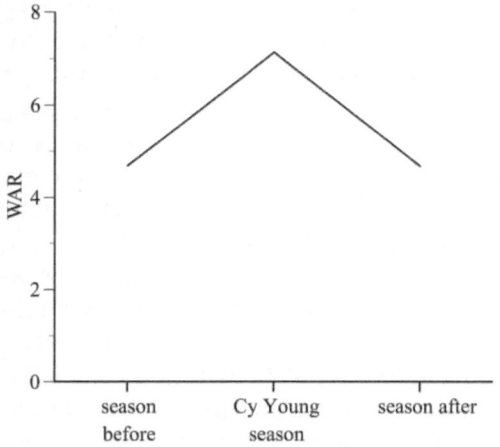

14.13 (This is a true story.) In the board game Settlers of Catan, a pair of six-sided dice are rolled on each turn to determine how new resources are distributed to each player. When Jim plays Settlers, he records every dice roll and inevitably finds that some numbers have come up more often than expected, while other numbers have come up less often than expected. For example, numbers 6 and 8 are equally likely, but they almost never come up equally often.

    a. Explain why numbers 6 and 8 are equally likely.

    b. Explain why you either are or are not surprised that numbers 6 and 8 seldom come up equally often.

    c. When a number that is advantageous for Jim (say, 8) comes up far less often than expected, he switches to a different pair of dice. Explain why you think that, going forward, 8 will or will not come up more frequently than it came up with the previous dice?

14.14 Thousands of pea seeds were divided into seven size categories and, for each category, the average size of the seeds of the offspring plants were recorded. A scatterplot compared the pea seed diameters (hundredths of an inch) of the parent plants and offspring plants. How would you explain the fact that the fitted line does not go through the origin with a slope of one?

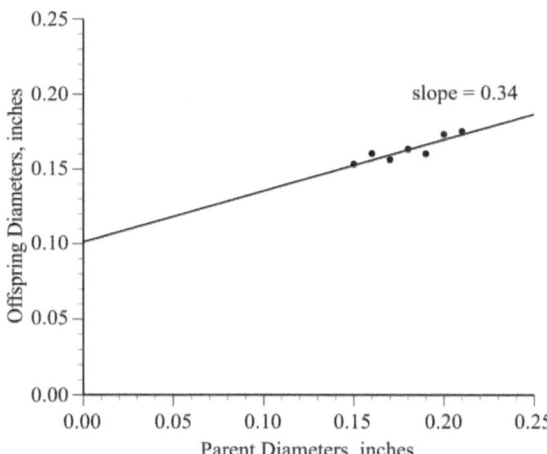

14.15 In speed climbing competitions, two climbers scramble up identical 15-meter walls, racing to be the first to touch a button at the top. Elite runner times average around 5 seconds. In tournament play, four climbers make it through the preliminary rounds to the semifinals. The winners of each of the two-person semifinal races advance to the "Big Final" where they compete for first and second place. The losers of the two semifinal races compete against each other in the "Small Final" for third and fourth place.

The first-, second-, and third-place finishers are "podium" places and win the majority of the prize money. A researcher hypothesized that the qualitative difference between third and fourth place incentivizes the climbers in the Small Final to exert more effort. His study of tournament data found that the two climbers competing for first and second place in the Big Final typically do not do as well as they did in their semifinal round while the two climbers competing for third and fourth place in the Small Final typically do better than they did in their semifinal round. Provide a purely statistical explanation for how this could happen even if each climber exerts the same amount of effort in each climb.

14.16 One hundred firms were grouped into quartiles based on their 1930 profits (return on assets): the top 25, second 25, third 25, and bottom 25. The average profits in 1930 and 1920 were then calculated for the firms in each of these 1930 quartiles. How would you explain the graph?

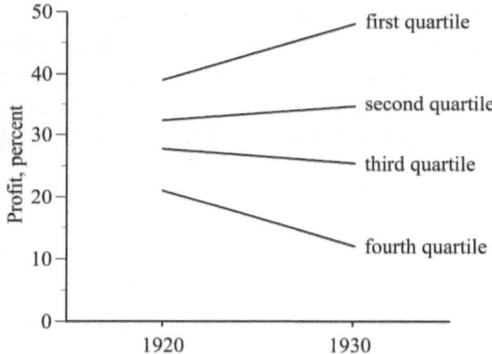

14.17 The John Bates Clark Medal is awarded annually to the top American economist under the age of 40. The 2021 medal was given to Isaiah Andrews. One of his interests is the "winner's curse," which was described by *The Economist* this way:

> when it comes to choosing between policies: the policy that performed best in a trial…will later be doomed to disappoint. To illustrate this the researchers turn to a trial that assesses the most effective ways of encouraging people to donate to charity, by combining requests for specific donations with promises to match the initial contribution. The researchers find that if the charity chooses the method that does best in a trial, it will always overestimate its donations.

Explain the winner's curse statistically.

14.18 A study of the 30 stocks in the Dow Jones Industrial Average over a 10-year period reported these estimates:

$$R = -3.25 + .56R_{-1} \quad R^2 = .587$$
$$(.46) \quad\quad (.14)$$

where:

R = firm's percent rate of return on assets that year

R_{-1} = firm's percent return on assets the previous year

Is there evidence of regression toward the mean? How do you know whether there is or isn't?

14.19 A statistics professor made a scatter plot of her students' midterm and final exam scores (each on a scale of 0–100), with midterm scores on the X-axis, and found that the least squares slope was less than 1.

a. How would you explain this?

b. If she had put final exam scores on the X-axis, do you think the slope would be less than or larger than 1?

14.20 How would you explain the broadcaster jinx?

Sports fans never like it when broadcasters bring up a lengthy positive streak belonging to their favorite team. Why jinx a free-throw shooter who's made 28 straight? A field goal kicker who hasn't missed from inside 40 the last three seasons? "There's no reason to tempt fate!" we yell from our couches and dens.

Considering that, fans of the Cleveland Browns had to cringe on Sunday when CBS put up a graphic noting their team hadn't turned the ball over in 99 trips to the red zone. They had to mutter when the broadcasting team of Andrew Catalon and Steve Beuerlein praised the Browns for their error-free ways. And they definitely had to be cursing as running back Isaiah Crowell immediately coughed up the ball just seconds later.

14.21 Use a specific example to explain why regression toward the mean is or is not the same as the law of averages.

14.22 For golfers who played the final two rounds of the 2015 and 2016 Masters golf tournaments, the correlation between their 2015 and 2016 scores was .38. For the top 15 golfers, the correlation between their 2015 and 2016 scores was .04. This is an example of

a. the law of averages.

b. the law of large numbers.

c. the central limit theorem.

d. self-selection bias.

e. the paradox of luck and skill.

14.23 What statistical pattern would you use to explain the fact that most movie sequels are not as good as the original?

14.24 Use one statistical argument to explain these two aphorisms: "The grass is always greener on the other side of the fence." "Familiarity breeds contempt."

14.25 Evergreen Day School, a selective suburban private school with a reputation for academic excellence, uses Educational Records Bureau (ERB) tests as an important component of the admissions process.

After admission to first grade, ERB tests are given each year to assess each student's progress. The following table uses percentiles to compare the ERB scores of students admitted to Evergreen in 2008 to the scores of suburban public schools. For example, the Evergreen first-graders scored, on average, in the 91.2 percentile compared to suburban public schools. The same Evergreen students took ERB tests as first, third, and fifth graders.

Grade	Reading Comprehension	Mathematics
1	91.2	90.2
3	67.9	73.2
5	66.1	72.5

How might a statistician explain these data, other than Evergreen is not as good as public schools? Explain your reasoning in words that the headmaster can understand.

14.26 A 2016 *Sports Illustrated* article noted that the Chicago Cubs had won 24 of their first 30 Major League Baseball games, and that only 13 teams had done so well in the modern baseball era. As a statistician, what pattern do you see in these data and how would you explain it?

	First 30 Games			Rest of Season		
	Wins	Losses	Percent	Wins	Losses	Percent
1905 Giants	24	6	80.0	81	42	65.9
1907 Cubs	24	6	80.0	83	39	68.0
1907 Giants	25	5	83.3	57	66	46.3
1911 Tigers	25	5	83.3	64	60	51.6
1921 Pirates	24	6	80.0	66	57	53.7
1928 Yankees	24	6	80.0	77	47	62.1
1939 Yankees	24	6	80.0	82	39	67.8
1946 Red Sox	24	6	80.0	80	44	64.5
1955 Dodgers	25	5	83.3	73	50	59.3
1958 Yankees	24	6	80.0	68	56	54.8
1977 Dodgers	24	6	80.0	74	58	56.1
1981 A's	24	6	80.0	40	39	50.6
1984 Tigers	26	4	86.7	78	54	59.1

14.27 Explain why you agree or disagree with this argument:

> *Regression toward the mean follows directly from the law of large numbers, which states that as the number of trials increases, the observed frequency with which an event occurs will converge to its probability. For example, in a large number of dice rolls, the above-average numbers (4, 5, and 6) will occur half the time and the below-average numbers (1, 2, and 3) will occur half the time. Thus, an above-average roll (like a 5) must be offset by a below-average roll (like a 2), which is regression toward the mean.*

14.28 In the 1930s, Horace Secrist, a statistics professor at the Northwestern University, wrote a book with the provocative title *The Triumph of Mediocrity in Business.* Secrist had found that businesses with exceptional profits in any given year tend to have smaller profits the following year, while firms with very low profits generally do somewhat better the next year. From this evidence he concluded that strong companies were getting weaker, and the weak stronger, so that soon all would be mediocre. How might a statistician explain these data, without making any economic assumptions?

14.29 A farmer sold some California land that she had been farming for 20 years and bought some farmland in the Midwest after investigating 20 different properties in five different states, spending one day at each property. Give a purely statistical explanation for why the property that appears to be the most attractive is probably not as good as it appears to be. Explain your reasoning carefully.

14.30 Marcus Lee suggested a betting strategy based on the logic of regression toward the mean and data on how professional football teams had done in recent weeks. Do you think he recommended betting for or against teams that had been doing poorly? Explain your reasoning.

14.31 Explain this observation by Nobel Laureate Daniel Kahneman: "As I understood clearly only when I taught statistics some years later, the idea that predictions should be less extreme than the information on which they are based is deeply counterintuitive."

14.32 In 2017, a famous economist argued that the ratio of the price of oil to the price of gold was a "golden constant":

> *The idea being that, if the price of oil changes dramatically, the oil-gold price ratio will change and move away from its long-term value. Forces will then be set in motion to move supply and demand so that the price of oil changes and the long-term oil-gold price ratio is reestablished. This represents nothing more than a reversion to the mean.*

Explain why you believe that this either is or is not an example of the regression-toward-the-mean principle.

14.33 Data were collected for golf scores on the first and second rounds of Professional Golf Association (PGA) tournaments for the years

2002–2018. The first-round scores in each tournament were sorted into better-than-average ("good") and worse-than-average ("bad"). The average round 1 and round 2 scores were then calculated for the good golfers and the bad golfers.

In the "Sort by Round 1" data in the table, D1 is the average difference between the round 1 scores of the Bad golfers and the Good golfers; D2 is the average difference between the round 2 scores of the round 1 Bad golfers and the round 1 Good golfers. (These are positive because low scores are good in golf.) The "Sort by Round 2" data in the table shows the same calculations when the golfers are sorted by how well they did in round 2.

Sort by Round 1		Sort by Round 2	
D1	D2	D1	D2
4.66	.72	.72	4.77

How do you explain the fact that D2 is less than D1 when the golfers are sorted by round 1 scores but the opposite is true when the data are sorted by round 2 performance?

14.34 A book and book review, both written by prominent economists, argued that the economic growth rates of nations converge over time. Milton Friedman wrote an apt commentary titled, "Do Old Fallacies Ever Die?"

> I find it surprising that the reviewer and the authors, all of whom are distinguished economists, thoroughly conversant with modern statistical methods, should have failed to recognize that they were guilty of the ... fallacy.... However, surprise may not be justified in light of the ubiquity of the fallacy both in popular discussion and in academic studies.

What fallacy is Friedman referring to?

14.35 Provide a purely statistical explanation for this observation: "Highly intelligent women tend to marry men who are less intelligent than they are."

14.36 In educational testing, a student's ability is defined as the student's (theoretical) average score on a large number of tests that are similar with respect to subject matter and difficulty. The student's score on a particular test is equally likely to be above or below the student's ability. Suppose that a group of 100 students takes two similar tests and the scores on each test have a mean of 65 with a standard deviation of 14. If a student's score on the second test is 52, do you predict that this student's score on the first test was: (a) below 52, (b) 52; (c) between 52 and 65; (d) 65; or (e) above 65? Explain your reasoning.

Answers

14.1 This is regression toward the mean.

14.2 This is regression toward the mean. Poor performances usually exaggerate how far a student's ability is below the mean. They will consequently generally do better on the next flight even if the instructors say nothing at all.

14.3 This is regression toward the mean. With randomly determined arrival times, extreme arrival times tend to be followed by arrival times closer to the mean.

14.4 Regression toward the mean does not assume that competition affects financial ratios. Regression is based on the fact that observed measurements that are far from the mean are probably farther from the mean than are the traits being measured.

14.5 This is the gambler's fallacy, not regression toward the mean.

14.6 No. If there is luck involved in famine and plenty, regression toward the mean predicts that an unusually large number of years of plenty will be followed by fewer years of plenty, not by an equally large number of years of famine.

14.7 This is regression toward the mean. A presidential-election winner is like the athlete drafted Number 1 or the person hired as CEO. Regression toward the mean teaches us that, like athletes and CEOs, the presidential candidates who win the election are seldom as good as they seem at the time.

14.8 This was a seminal study of regression toward the mean. There is inevitable variation in heights, even among the children of the same mother and father. An unusually tall person probably had somewhat shorter parents and will probably have somewhat shorter children even while the variation in heights is stable.

14.9 The logic underlying the regression-toward-the-mean phenomenon suggests that he is the tallest. The height of someone who is very tall is probably an overstatement of this person's genetically inherited height—which is also the genetically inherited height of the brothers.

14.10 Regression toward the mean is a compelling statistical argument. Those individuals or teams that appear on the cover of *Sports Illustrated* are not a random sample. They have typically done something exceptional recently—perhaps won the World Series, a major tennis tournament, or 47 football games in a row. Such accomplishments are more likely to have involved good luck than bad and are not likely to continue indefinitely.

14.11 This is regression toward the mean.

14.12 This is regression toward the mean. The most successful players in any given year tend to have had more than their share of good luck, which cannot be counted on to persist.

14.13 a. There are five ways to roll a 6 (1-5, 2-4, 3-3, 4-2, and 5-1) and five ways to roll an 8 (2-6, 3-5, 4-4, 5-3, and 6-2), so each has a probability 5/36.

　　 b. In a small number of rolls, it is unlikely that every observed frequency will be close to the theoretical probability.

　　 c. When something has been happening far less often than expected, it is likely that it will, on average, come up as often as expected in the future—which is more frequently than it has been coming up.

14.14 This is regression toward the mean—just like the heights of the children of unusually tall or short parents tend to regress toward the mean.

14.15 This is regression toward the mean. For any given amount of effort, each climb inevitably has small variations in the climbing time. Those who make it to the Big Final are more likely to have had beneficial variations in the semifinal round—and can be expected to do somewhat worse in the Big Final—while those who compete in the Small Final are more likely to have had detrimental variations in the semifinal round— and can be expected to do somewhat better in the Small Final.

14.16 A firm's profits fluctuate. There is regression toward the mean because firms with observed profits that are far from the overall mean tend to have average profits that are closer to the overall mean. Thus, their profits in any other year will generally be closer to the mean.

14.17 This is regression toward the mean. Since there is randomness in the results, the policy that performed the best is probably not as far above average as suggested by the trial.

14.18 The coefficient of R_{-1} is less than 1, so if the return is X percentage points higher than average one year, it is predicted to be only .56X percentage points higher than average the next year.

14.19 a. This is regression toward the mean. Those with the most extreme scores on the midterm tend to score closer to the mean on the final.

　　 b. Less than 1, again because of regression toward the mean. Those with the most extreme scores on the final tend to score closer to the mean on the midterm.

14.20 This is regression toward the mean plus selective recall. No team has a 0% probability of fumbling, so a 0% fumble rate exaggerates their ability to not fumble. We remember when announcers are immediately contradicted and forget the times when they are not contradicted.

14.21 They are different. Suppose a student gets a very high score on the midterm. Regression toward the mean implies that this student's score on the final exam is likely to be closer to the average score. The fallacious law of averages says that this student's score on the final exam will likely be below average.

14.22 This is the paradox of luck and skill. In contests among the most skilled, the outcomes are usually determined by luck.

14.23 This is regression toward the mean. There is considerable uncertainty about whether a movie will be a success, and sequels are made of the most successful movies, which probably had more good fortune than bad. (If you want sequels to be better than the original, make sequels of unsuccessful movies.)

14.24 This is regression toward the mean.

14.25 Regression toward the mean is a possible explanation. Those students whose high first-grade scores qualified them for admission to Evergreen probably had abilities that were somewhat closer to the mean. When they took the ERB tests again as third and fifth graders, their scores reflected their more modest ability. (The name of the school has been changed, but these are real data.)

14.26 This is clear evidence of regression toward the mean. These 13 teams generally won more than half their games the remainder of the season (they were above-average teams), but none of them did as well after the first 30 games as they did during the first 30 games.

14.27 The quotation is describing the fallacious law of averages. Regression toward the mean says that an outcome that is far from average (like a 1 or 6) will probably be followed by something closer to average.

14.28 The president of the American Statistical Association wrote an enthusiastic review of this book; another statistician pointed out that Secrist had been fooled by regression toward the mean. In any given year, companies with exceptional profits relative to other companies are likely to have experienced good fortune.

14.29 This is an example of regression toward the mean. Some properties will look more attractive than they really are and some will look less attractive than they really are. The property that looks the most attractive is more likely to be in the former group than in the latter.

14.30 The performance of a professional football team depends on its ability and also on chance—unpredictable fluctuations in the players' health, the officiating, and even the proverbial bounces of the football. When such chance events make observed performance an imperfect measure of ability, observed differences in performance tend to exaggerate differences in abilities. Performances consequently regress toward the mean, in that those teams that perform the best or worst at any point in time typically do not perform as well or poorly subsequently. Lee recommended betting on teams that had been doing poorly and betting against teams that had been doing well, and it worked.

14.31 He was referring to regression toward the mean. When something (like ability) is measured imperfectly, observations that are far from the mean generally reflect traits that are closer to the mean.

14.32 Disagree. The regression-toward-the-mean principle says that when something is measured imperfectly, measurements that are far from the mean overstate how far the trait being measured is from the mean; for example, the highest score on a test probability overstates how far the student's ability is from the mean. This principle is important but it has nothing to do with the ratio of the price of oil to the price of gold.

14.33 This is regression toward the mean. No matter whether we sort golfers by their performance in round 1 or round 2, the best performing golfers typically benefitted from some good luck and will not do so well, on average, in the other round.

14.34 They completely ignored the role of regression toward the mean in this convergence.

14.35 Most men are less intelligent than a highly intelligent woman. Unless the intelligence of spouses is perfectly correlated, so that nothing but intelligence matters in match-making, highly intelligent women will, on average, marry men who are of more average intelligence.

14.36 Regression toward the mean teaches us that this student's ability is probably below average, but not as far below average as was this test score. The score on the second test is most likely between 52 and 65.

15

Multiple Regression

15.1 This multiple regression equation was estimated using 2005 data for 100 homes in Plano, Texas:

$$Y = 15.034 + 104.96S + 56.68G - 12{,}018A - 5{,}225BR + 3{,}396BA + 23{,}835AC, \; R^2 = .774$$
$$(2.206) \quad (11.70) \quad (13.70) \quad (793) \quad (2{,}187) \quad (1{,}235) \quad (4{,}793)$$

where:

  Y = home price
  S = square feet of living area
  G = garage square feet
  A = age of house in years
  BR = number of bedrooms
  BA = number of bathrooms
  AC = 1 if it has central air, 0 if not

and standard errors are in parentheses.

a. Which of the estimated coefficients are statistically persuasive?

b. Provide a logical explanation for why the coefficient of BR might be negative.

c. Do you think that the coefficients in housing regression equations change over time? Why or why not?

15.2 Sales data for single-family houses in an Indianapolis suburb were used to estimate this equation in 2005:

$$Y = 60.076 + 75.1S + 36.4G - 3{,}295A + 4{,}473B - 14.632T, \; R^2 = .84$$
$$(14.4) \quad (19.7) \quad (12.1) \quad (1{,}076) \quad 1{,}708 \quad 3{,}531$$

where:

  Y = sale price
  S = square feet of living area
  G = garage square feet
  A = age of house in years
  B = number of baths
  T = 1 if it is two-story, 0 if not.

DOI: 10.1201/9781003630159-15

The standard errors are in parentheses.

a. Which of the estimated coefficients are statistically significant at the 5% level? How can you tell?

b. Does the sign and size of each coefficient seem reasonable? In each case, explain your reasoning.

15.3 Identify five errors in these reported results from a multiple regression model for predicting the number of home sales, using quarterly time-series data from 2001 to 2020:

Explanatory Variable	Coefficient	Standard Error	|t-value|	Two-sided p-value
Intercept	−1703.00	412.00	4.10	.0001
GDP	32.33	6.20	5.21	−.0271
Mortgage rate	−2245.61	711.42	3.16	.0023
Spring	45.62	28.62	1.59	.1161
Summer	35.55	21.44	1.66	.1012
Fall	−12.11	4.63	2.62	.9894
Winter	−48.79	15.72	3.10	.0027

$n = 20$, $R^2 = 1.69\%$

where:
  spring = 1 if spring, 0 otherwise
  summer = 1 if summer, 0 otherwise
  fall = 1 if fall, 0 otherwise
  winter = 1 if winter, 0 otherwise

15.4 Identify five distinct problems with this multiple regression model of the effect of COVID-19 on the earnings of small business owners.

$$\text{Earnings} = \alpha + \beta_1 \text{Covid} + \beta_2 \text{Race} + \varepsilon$$

Earnings = annual income from business, dollars
COVID = 1 during COVID months, 0 pre-COVID
Race = 0 if white, 1 if black, 2 if Asian

Variable	Coefficient	Standard Error	t-statistic	Two-sided p-value
Constant	34,256.77	2,437.92	14.05	.0385
COVID	1,327.38	770.95	1.72	.0427
Race	450.22	125.43	.28	.000165

$\varepsilon = -2.21$; $R^2 = .10$.

15.5 A study of the income of family-law lawyers looked at the influence of age, sex, and law school attended:

$$Y = \alpha + \beta_1 X + \beta_2 D_1 + \beta_3 D_2 + \varepsilon$$

where:
  Y = average income over the past 5 years
  X = current age
  D_1 = 0 of male, 1 if female
  D_2 = 1 if attended Harvard, 2 if attended Yale, 3 if attended Stanford, and 0 otherwise.

What is the most important problem with this model specification?

15.6 A student estimated a regression model that used the number of pages in best-selling hardcover books to predict their prices. Explain why this conclusion is wrong: "The 0.75 R-squared is pretty large, when you consider that the price of a book is $20 and .75($20) = $15 of the price is explained by the number of pages."

15.7 A researcher argued that, "Multicollinearity inflates the standard errors of the coefficients of the explanatory variables, so that there is a higher likelihood of rejecting the null hypothesis for these coefficients" Why is he wrong?

15.8 A researcher used data for 60 randomly selected colleges: Y = percent graduation rate (mean 59.65); X_1 = students' median math plus reading/writing SAT score, (mean 1,030.5); and X_2 = percent of students with GPAs among the top 10% at their high school (mean = 46.6). He found a statistically significant relationship between graduation rate and SAT scores (the t-values are in brackets):

$$Y = -33.39 + .090X_1, \ R^2 = .71$$
$$[2.34] \qquad [6.58]$$

and between the graduation rate and GPA:

$$Y = 42.20 + .375X_2, \ R^2 = .52$$
$$[8.93] \qquad [4.46]$$

However, when he included both SAT scores and GPAs in a multiple regression equation, the estimated effect of GPA on graduation rates was very small and not statistically significant at the 5% level:

$$Y = -26.20 + .081X_1 + .056X_2, \ R^2 = .71$$
$$[1.24] \qquad [3.30] \qquad [.47]$$

a.   He suspected that there was an error in his multiple regression results. Is it possible for a variable to be statistically significant in a simple regression, but not significant in a multiple regression?

b.   Do you think that X_1 and X_2 are positively correlated, negatively correlated, or uncorrelated?

c.   If your reasoning in (b) is correct, how would this explain the fact that the coefficients of X_1 and X_2 are each lower in the multiple regression equation than in the simple regression equations?

15.9   Suppose that a college student's first-year grade point average G can be predicted from the student's high school grade point average H and standardized test scores T: $G = \alpha + \beta_1 H + \beta_2 T + \varepsilon$, where β_1 and β_2 are positive, and H and T are positively correlated. If T is omitted from this equation, how will least squares estimate of the coefficient of H be affected? What harm is there if T is omitted?

15.10   A study compared the daily returns for a portfolio of stocks that have clever ticker symbols to the performance of the overall market for the years 1984–2005. Part of the study involved the estimation of this regression equation,

$$Y = .00049 + .81X, \quad R^2 = .29$$
$$[3.45] \quad [39.52]$$

where Y is the clever-ticker daily return, X is the stock market daily return, and the t-values are in brackets.

An update of this study using data for 2006–2018:

$$Y = .00026 + .88X, \quad R^2 = .68$$
$$[1.78] \quad [64.56]$$

How could you test if the differences in the estimated coefficients (.00049 versus .00026 and .81 versus .88) are statistically persuasive? Be specific.

15.11   A study of the effect of birth order on GPA reported these results:

$$Y = 4.24 + .025X1 - .45X2 + .56X3 + .25X4 - .88X5, \quad R^2 = .47$$
$$[3.11] \quad [2.05] \quad 1.67 \quad [2.79] \quad [1.98] \quad [2.22]$$

where:
   Y = GPA, scale of 0–4
   X1 = height, in inches
   X2 = 1 if female, 0 otherwise
   X3 = 1 if only child, 0 otherwise

X4 = 1 if first born, 0 otherwise
X5 = 1 if not an only child or first-born, 0 otherwise
[ ] = absolute value of t-values

Why do you suspect that this is not an honest report of the results?

15.12 In a 1980 court case, a federal judge described how regression estimates of male and female salary equations could be used to determine if an employer discriminates against females.

$$\text{Males: } Y = \alpha_M + \beta_M S + \gamma_M E$$

$$\text{Females: } Y = \alpha_F + \beta_F S + \gamma_F E$$

where:
Y = annual earnings
S = number of years of schooling
E = number of years of relevant job experience

The judge also noted that in place of two separate equations, one for males and one for females, the following single equation could be estimated, where D = 0 if the person is a male and D = 1 if the person is female:

$$Y = \alpha + \beta_1 S + \beta_2 E + \beta_3 D$$

a. In what ways is this single equation less general than the two separate equations?
b. What single equation could the judge have specified that is as general as the two separate equations?

15.13 The following multiple regression model was used to explain the sale prices (in thousands of dollars) for 60 single-family homes in Oxnard, California:

Variable	Coefficient	Standard Error
Constant	−466.2559	164.7577
Distance to waterfront, meters	−.1515	.0681
Living area, square feet	.7670	.2870
Number of bedrooms	−358.5812	124.7850
Number of bathrooms	468.5877	269.7153
$R^2 = .682$		

a. Interpret the −.1515 coefficient.
b. Which coefficient estimates (other than the constant) are statistically persuasive?

c. Explain why you either agree or disagree with this argument: "The bedroom coefficient is negative because if the living area is fixed, doubling the number of bedrooms halves the size of each."

d. Explain why you either agree or disagree with this argument: "The bathroom coefficient is positive because homes with more bathrooms tend to be larger than homes with fewer bathrooms."

15.14 A study of home prices in a suburban Rhode Island town used these variables in a multiple regression model:

Y = price

X_1 = square feet

X_2 = number of bedrooms

X_3 = number of bathrooms

X_4 = Walk Score, a rating (from 0 to 100) for a given address, based on the walking distance from that address to a variety of key amenities. A Walk Score of 70 or above is generally considered good.

What is wrong with this interpretation of the results: "The coefficient of X_4 is negative because large homes that have high prices are usually secluded from grocery stores, coffee shops, and other desirable destinations."

15.15 Here is a table of descriptive statistics for data that were used to see whether the income of a company's workers depend on sex or race (Sex = 1 if male, 0 otherwise; Race = 1 if white, 0 otherwise). Put these five numbers in the appropriate places: 0, .60, 1, 37897, 24246:

Variable	Mean	Median	Standard Deviation	Minimum	Maximum
Income	33,830	???	27,846	2,343	130,000
Sex	???	0	.50	0	1
Race	.25	???	.72	0	???
N = ???					

15.16 Identify two ways in which this report of the estimation of a multiple regression equation could be improved:

Variable	t-statistic	p-value
Variable1	2.47	.02
Variable2	1.70	.09
Variable3	6.31	<.001

15.17 Traffic fines for speeding in California are supposed to be determined by a formula based on how many miles over the speed limit the driver was driving, but there is flexibility in various extra fees. A researcher wants to see if total speeding fines are higher for people under the age of 30 than for older drivers.

a. Specify a multiple regression equation that might be used to investigate this research question.

b. Which coefficients in your model would show whether there is discrimination against younger drivers?

c. Identify a situation in which a difference-in-means test would show discrimination against younger drivers, but a multiple regression model would not.

d. Identify a situation in which a difference-in-means test would show no discrimination against younger drivers, but a multiple regression model would.

15.18 A dean claims that humanities professors' salaries Y are determined solely by teaching experience X. To see if there is a statistically persuasive difference in the salaries of male and female humanities professors, the equation $Y = \alpha + \beta_1 D + \beta_2 X + \beta_3 DX + \varepsilon$ was estimated, using the variable $D = 0$ if male, 1 if female.

a. Interpret each of the parameters α, β_1, β_2, and β_3.

b. What advantage does a regression model have over a comparison of average male and female salaries?

c. Describe a specific situation in which a comparison of average male and female salaries shows discrimination against females while a regression equation does not.

d. Describe a specific situation in which a comparison of average male and female salaries shows no discrimination against females while a regression equation indicates discrimination against females.

15.19 A researcher was interested in how a movie's box office revenue is affected by the audience suitability rating assigned by the Motion Picture Association of America (MPAA) and the quality of the film as gauged by Entertainment Weekly's survey of ten prominent film critics. Data for 105 films that were critically rated in 1991 were used to estimate the following equation by ordinary least squares:

$$Y = 6,800,024 - 25,232,790D + 6,453,431X, \quad R^2 = .214$$
$$\quad [.61] \qquad\qquad [4.36] \qquad\qquad [3.74]$$

where:

Y = domestic box office revenue, in dollars

D = MPAA suitability rating (D = 0 if rated G, PG, or PG-13; D = 1 if rated R)

X = Entertainment Weekly quality rating (12 = A, 11 = A–, 10 = B+, ...)

[ ]: t-values

Explain the errors in each of these three critiques:

a. "The R-squared value of 0.214 is close enough to zero that the null hypothesis can be rejected."

b. "The coefficient of X has the wrong sign and is also much too large."

c. "The low t-value of the alpha value suggests that the y-intercept may not be significant. Therefore, while this may be a good study of films in 1991, it may not be a good study for predicting films."

15.20 A student estimated a regression model using annual data for 1990 through 2015,

$$C = \beta_0 + \beta_1 Y + \beta_2 W + \varepsilon$$

where:
C = consumption spending
Y = disposable income
W = household wealth, all in billions of dollars

The initial estimate of β_2 was not statistically significant at the 5% level; so the student omitted wealth from the equation, arguing that, "There could have been multicollinearity between income and wealth. Therefore, the regression is more accurate without wealth included." Explain why you either agree or disagree with that reasoning.

15.21 How would you respond to this question?

> *I have run into a problem with missing values in my data. Some of the variables including GPA and parents' education have a lot of missing values. If I discard those individuals with missing values, then the dataset becomes very small, but if I replace all the missing values with zeros, the total sample size is about 35,000 and I am able to get statistically significant results.*

15.22 Identify the most serious problem with this model of household income

$$Y = \alpha + \beta_1 D + \beta_2 E + \beta_3 A + \varepsilon$$

where:
Y = average household income in state
D = state dummy variable; 1 for Alabama, 2 for Alaska, 3 for Arizona, etc.
E = average education in state
A = average age in state

15.23 This model was estimated by ordinary least squares using 2011 data for individuals:

$$Y = \alpha + \beta_1 A + \beta_2 E + \beta_3 F + \beta_4 M + \beta_5 U + \varepsilon$$

where:

   Y = logarithm of income
   A = age
   E = education, years
   F = female; 1 if female, 0 if male
   M = married; 1 if married, 0 if unmarried
   U = unmarried; 0 if unmarried, 1 if married

What statistical problem do you see?

15.24 In 1960, Meyer Efroymson, a statistician for Esso Research and Engineering, proposed a stepwise procedure for choosing the explanatory variables in a multiple regression model from a list of candidate variables, one by one, based on which variable has the lowest p-value. One explanatory variable is chosen in the first round; a second explanatory variable is added in the second round, and so on, until there are no remaining candidate variables with p-value less than .05. This procedure was recently tested using 100 observations on 50 candidate variables. The estimated model was then used to make predictions for 100 additional observations. The mean square error of the predictions was 348.20 for the 100 observations used to estimate the model and 2,738.43 for the 100 new observations. How would you explain this disparity?

15.25 A researcher commented on the estimated equation shown below: "Multicollinearity could be a factor, as indicated by the very high standard error and the small *t*-value."

$$Y = 1{,}146.0 + 1.432X, \quad R^2 = .00$$
$$ (195.6) \quad\quad (71.6)$$
$$ [5.86] \quad\quad [.02]$$

( ): standard errors

[ ]: t-values

a. How does multicollinearity affect standard errors and t-values?

b. Why is it clear that multicollinearity is not a problem here?

15.26 This regression equation was estimated using data for 481 California Census blocks that were more than 95% Hispanic in 2008:

$$P = .843 + .00023Y + .00038D, \quad R^2 = .0025$$
$$ [97.12] \quad [1.11] \quad\quad [8.42]$$

where:

P = % of the two-party vote received by Barack Obama in the 2008
Presidential election

Y = median household income of Hispanics ($1000s of dollars)

D = 0 if the Census block is in L.A. County; D = 1 otherwise

and the t-values are in brackets. Explain why you either agree or disagree with these conclusions:

a. Income is not statistically persuasive; the p-value is .27, meaning that there's a 27% chance we accept the null hypothesis that income has no effect on the Hispanic vote.

b. The effect of D on P is substantial because the t-value is 8.42.

c. There is an error in the results because the R^2 cannot be so low if some explanatory variables are statistically significant.

d. The model would be better if D = 1 if the Census block is in Los Angeles County and D = 0 otherwise.

15.27 A random sample of 100 college students yielded data on Y = student's height, X_1 = student's sex (0 if female, 1 if male), X_2 = mother's height, and X_3 = father's height. All heights were measured in feet; for example, a 5 foot, 3-inch person is 5.25 feet tall.

$$Y = .821 + .434X_1 + .364X_2 + .456X_3, \quad R^2 = .74$$
$$(.511) \quad (.038) \quad (.087) \quad (.077)$$
$$[1.61] \quad [11.31] \quad [4.19] \quad [5.91]$$

a. Explain why the following interpretation of the .74 R^2 value is misleading: "74% of a student's height can be explained by their sex and their parents' height."

b. Explain why this conclusion is misleading: "The student's sex is important in determining height. In general, males are taller than females, which is shown in the t-values."

c. Do these results indicate that heights regress toward the mean?

15.28 An equation was estimated to gauge the effect of the unemployment rate (X_1) and the Treasury bond rate (X_2) on stock prices (Y):

$$\hat{Y} = 502.50 - 12.12X_1 - 25.85X_2, \quad R^2 = .904$$
$$(42.64) \quad (4.17) \quad (5.03)$$

( ): standard errors

During the period studied, the Federal Reserve used high interest rates to weaken the economy (and reduce inflation) and then used low interest rates to stimulate the economy. As a consequence, the unemployment rate and interest rate were highly positively correlated. To cure this multicollinearity problem, the researcher omitted the interest rate and obtained this result:

$$\hat{Y} = 482.77 - 65.09X_1, \ R^2 = .702$$
$$(39.85) \qquad (3.60)$$

Explain why the cure is worse than the problem.

15.29 A blogger for the statistical software program Minitab answered the question, "How Can I Deal with Multicollinearity?" by writing that "the solution may be relatively simple.... Remove highly correlated explanatory variables from the model.... Because they supply redundant information, removing one of the correlated factors usually doesn't drastically reduce the R-squared." How would you respond?

15.30 Suppose that the percentage of the popular vote for US President received by the incumbent party's candidate is related negatively to both the unemployment rate and the rate of inflation. A student learned in a macroeconomics class that there is a negative correlation between unemployment and inflation, and consequently omitted the rate of inflation in order to avoid a multicollinearity problem:

$$V = \alpha + \beta U + \varepsilon$$

Predict the effect of this omission on the estimated coefficient of the unemployment rate. Explain your reasoning so that a novice will understand your argument.

15.31 A researcher used 2011 data for 30 developing countries to estimate a model of real per capita GDP, in US dollars. The six education variables are the percent of people of the appropriate age enrolled in school; for example, the percent of females of primary school age who are enrolled in primary school. The adolescent fertility rate is births per 1,000 women ages 15–19.

The researcher initially estimated Model 1. After noting the high multicollinearity among the explanatory variables, the researcher estimated Model 2.

	Model 1		Model 2	
	Coeff	t-value	Coeff	t-value
Intercept	92,229	1.17	16,380	.72
Female primary	1,154	.64	726	1.53
Male primary	−2,219	.96	−899	1.68
Female secondary	1,771	1.57	−1,081	2.21
Male secondary	1,901	1.80	1,113	2.16
Female tertiary	265	.47	263	.71
Male tertiary	−84	.10	192	.37
Volume of exports	429	.11		
Unemployment rate	−290	.17		
Adolescent fertility rate	−89	.43		
R-squared	.804	.684		

The author concluded that, "The R-squared decreased from 0.804 to 0.684. However, this model provides more precise estimates of the coefficients of the explanatory variables. The standard errors fall by almost half and the t-values for almost all variables, specifically secondary education for both sexes, are statistically significant, unlike in the previous model." What do you think?

15.32 In an IBM antitrust case, an economist, Franklin Fisher, estimated multiple regression equations explaining computer prices as a function of memory, speed, and other characteristics. The t-values for all of the regression coefficients were 20 or higher. However, a government expert, Alan K. McAdams, argued that multicollinearity among the explanatory variables made it impossible to draw reliable conclusions about the effects of the different explanatory variables on computer prices. Explain why you either agree or disagree with McAdams' logic.

15.33 A multiple regression model was used to estimate the relationship between a Major League Baseball (MLB) player's salary and his batting average and home runs:

	Mean	Standard Deviation
S = annual salary, millions of dollars	2.5	3.5
B = batting average for season	.263	.032
H = number of home runs in season	12.5	10.4

	Coefficient	Standard Error	t-value
Constant	−11.70	.97	12.1
B	13.24	.98	13.5
H	.15	.0031	48.4
$R^2 = .26$			

Can we conclude from these results that salaries are affected more by batting average than by home runs, or vice versa? Explain your reasoning.

15.34 Interpret these regression results using quarterly data for 1974–2014. The dependent variable is housing starts, in thousands. The explanatory variables are:

Time = 1 in the first quarter of 1974, 2 in the second quarter of 1974, etc.

Q2 = 1 in second quarter, 0 otherwise

Q3 = 1 in third quarter, 0 otherwise

Q4 = 1 in fourth quarter, 0 otherwise

Variable	Coefficient	Standard Error	t-value	One-Sided P
Intercept	2,806.65	1,090.71	2.57	.0055
Time	−1.30	.55	2.37	.0094
Q2	90.93	18.14	5.01	.0000
Q3	70.97	18.14	3.91	.0001
Q4	19.25	18.25	1.05	.1466
SEE = 82.13; $R^2 = .196$				

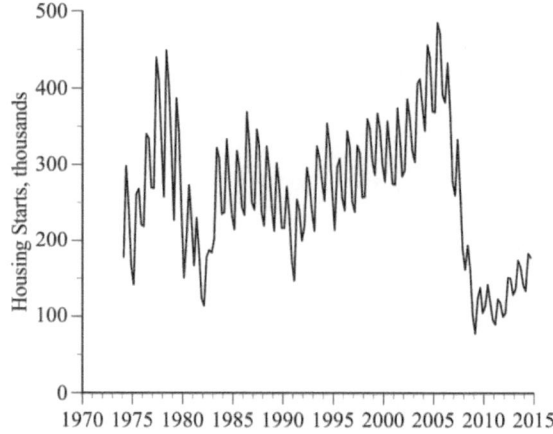

15.35 Data for the top 100 money winners on the 1993 Professional Golf Association (PGA) tour were used to estimate the following equation:

$$Y = -8,023,111 + 15,992D + 12,003A + 27,003G + 91,306P + 11,357S$$

   (1,103,835)   (3,547)   (6,571)   (10,078)   (19,649)   (3,705)
     [7.26]   [4.51]   [1.83]   [2.69]   [4.65]   [3.07]

where:
   Y = dollars won
   D = average length of drives from the tee, yards
   A = driving accuracy (percent of drives landing in fairway)
   G = percent of greens that were reached in par minus two strokes
   P = percent of those greens reached in par minus two that required only one putt
   S = percent of times player made par after hitting into sand

The standard errors are in parentheses; the t-values are in brackets; and $R^2 = .468$.

a. Which of the explanatory-variable coefficient are statistically significant at the 5% level?

b. Which, if any, of the estimated explanatory-variable coefficients have plausible signs?

c. How can the intercept be negative if Y cannot be negative?

d. Does the fact that the coefficient of P is larger than the coefficient of D imply that, other things being equal, a player who is a good putter will win more money than a person who hits long drives? Explain.

e. Does the fact that the t-value for the coefficient of D is larger than the t-value for the coefficient of S imply that, other things being equal, a player who hits long drives will win more money than a person who is better at making par out of sand traps? Explain.

15.36 A multiple regression equation was used to explain rents paid by people who live and work in Phoenix:

$$Y = \alpha + \beta_1 X + \beta_2 D + \varepsilon$$

where:
   Y = monthly rent
   X = monthly income
   D = 1 if full-time job, 0 if part-time

Explain why you either agree or disagree with this interpretation of β_2: "The coefficient of the dummy variable measures the extent to which renters can afford to pay a higher rent because their full-time job gives them more income."

15.37 A survey of 47 college sophomores investigated the effect of studying and extracurricular activities on grades:

$$Y = 8.809 + .069X - .085D$$
$$(.339) \quad (.017) \quad (.289)$$

where:

Y = grade point average on a 12-point scale
X = average hours per week spent studying;
D = 1 if the person spends at least 10 hours a week on an extracurricular activity such as work or sports, 0 otherwise

and the standard errors are in parentheses. The researcher concluded that, "The effect of extracurricular activity is not significant and does not lend support to the commonly held notion that extracurricular activity negatively affects grades because it reduces available study time." Explain why the coefficient of D does not measure the extent to which extracurricular activity affects grades by reducing available study time.

15.38 A researcher specified this time-series model of consumer spending,

$$C = \alpha + \beta_1 Y + \beta_2 D + \varepsilon$$

where:

C = consumer spending
Y = disposable income
D = 1 if the stock market went up that year, 0 otherwise

What errors do you see in these reported results?

a. t = 3.64 for a test of H_0: Y = 0.

b. The value of β_2 should be positive because when income goes up, the stock market usually goes up too.

c. For β_2, t = 2.01 and the two-sided p-value is .95.

d. $R^2 = 1.95$.

e. The estimate of ε is 1.13.

15.39 What is the most important thing wrong with these multiple regression results that examined the relationship between year in college and mental depression?

Dependent variable:

Depression = survey, measured on a scale of 0–10

Explanatory variables:

Frosh = 1 if first year, 0 otherwise

Sophomore = 1 if second year, 0 otherwise

Junior = 1 if third year, 0 otherwise

Senior = 1 if fourth year or longer, 0 otherwise

Results

Constant	4.720***
	(1.664)
Frosh	.669
	(.818)
Sophomore	.802
	(.850)
Junior	.923
	(.838)
Senior	.819
	(.814)
Observations	113
R-Squared	.531

Standard errors in parentheses
*** $p < .01$, ** $p < .05$, * $p < .10$

15.40 What are the problems with these statements?

    a. "I have data on 11 possible explanatory variables for my multiple regression model. I will use the three variables with the highest t-values."

    b. The high R-squared shows that all of these explanatory variables have large effects on Y.

    c. "Since the p-value for the coefficient of income is 0.0000000427, we are very confident that there is less than a 5% probability that the null hypothesis is true."

    d. "There is a multicollinearity problem: The explanatory variables are correlated with each other but the multiple regression model assumes that the explanatory variables are independent."

Answers

15.1 a. All of them since the t-values (estimate divided by standard error) are all much larger than 2.

   b. The coefficient of BR might be negative because it is *ceteris paribus*, holding the values of the other explanatory variables constant. Holding square footage constant, people may not want (for example) an extra bedroom at the cost of a smaller living room, or smaller bathrooms.

   c. The parameters of housing regression equations change over time because construction costs, land values, and tastes change.

15.2 a. All the estimated coefficients are statistically significant at the 5% level because all the t-values (estimate divided by standard error) are much larger than 2.

   b. They all seem reasonable. An extra square foot of living area is predicted to increase the sales price by $75.1. An extra square foot of garage is predicted to increase the sales price by $36.4. An extra year of age is predicted to reduce the price by $3,295. An extra bathroom is predicted to increase the sales price by $4,473. Being two-story *ceteris paribus* (holding total square footage constant) is predicted to reduce the price by $14,632; evidently people in this area prefer their square footage to be on one level.

15.3 Five errors: (1) with 4 seasons and 20 years, there are 80 observations, not 20; (2) R^2 is not a percent but a number from 0 to 1; (3) p-values can't be negative; (4) for t = 2.62, the p-value is much lower than .9894; and (5) with four seasonal categories, there should be only three dummy variables. (In fact, the equation cannot be estimated as specified.)

15.4 a. The Race variable should be two 0-1 dummy variables.

   b. The two-sided p-value for the constant should be much lower than .05. (The correct two-sided p-value is 3.85×10^{-14}.)

   c. The two-sided p-value for COVID should not be lower than .05. (.0427 is the one-sided p-value.)

   d. The t-statistic for Race should be larger than 1. (The reported number is SD/b instead of b/SD.)

   e. There is no single estimate for ε.

15.5 This specification for D_2 assumes an unrealistic linear relationship, with the effect of law school attended on salary being twice as big for Yale as for Harvard, and three times as big for Stanford as for Harvard. With four school categories, there should be three 0–1 law school dummy variables; for example,

$D_2 = 1$ if attended Harvard, 0 otherwise

$D_3 = 1$ if attended Yale, 0 otherwise

$D_4 = 1$ if attended Stanford, 0 otherwise

15.6 R-squared = .75 means that 75% of the *variation* in the prices is explained by the number of pages.

15.7 A higher standard error reduces the t-value, reducing the likelihood of rejecting the null hypothesis.

15.8 The first two equations show positive statistically persuasive relationships between SAT scores and graduation rates and between GPAs and graduation rates. These positive effects persist in a multiple regression equation that includes both explanatory variables, although the SAT coefficient and t-value drop slightly and the GPA coefficient drops dramatically and is no longer statistically significant at the 5% level. Evidently, SAT scores and grades are positively correlated, so that the omission of either from the multiple regression equation biases the estimated coefficient of the included variable upward. (The combined positive effects of both variables are attributed to a single variable.)

a. Yes, the omission of an important explanatory variable that is correlated with an included explanatory variable can bias the estimated coefficient of the included variable, thereby making it statistically significant.

b. SAT scores and high school GPAs are most likely positively correlated.

c. If the multiple regression equation is correct, so that SAT scores and high school GPAs both have positive effects on a school's graduation rate, and these two explanatory variables are positively correlated, then the omission of either variable from the regression equation makes it appear that the other variable has a larger effect.

15.9 When GPA (H) increases, test scores (T) tend to increase, too, and both have a positive effect on first-year GPA. If test scores are omitted from the estimated equation, then it will appear that high school GPA has an exaggerated effect on first-year GPA. Therefore, if T is omitted, the estimated coefficient of H will be biased upward. In addition, the model will not make accurate predictions.

15.10 We can use a dummy variable for the different time periods. For example, D = 1 if 2006–2018, and D = 0 if 1984–2005, and

$$Y = \alpha + \beta_1 X + \beta_2 D + \beta_3 DX + \varepsilon$$

The p-value for β_2 tests whether the intercepts are the same and the p-value for β_3 tests whether the coefficients of X are the same.

15.11 There should only be two dummy variables for the three birth-order categories. The equation cannot be estimated with these three dummy variables.

15.12 a. The judge's equation forces the male and female equations to have the same coefficients of S and E.

b. This equation gives the same results as the two separate equations and, in addition, allows us to test whether the differences in the male and female coefficients of S and E are statistically persuasive:

$$Y = \alpha + \beta_1 S + \beta_2 E + \beta_3 D + \beta_4 DxS + \beta_5 DxE$$

15.13 a. Each 1 meter increase in distance from water is predicted to reduce the sale price by \$151.50.

b. Each t-value is equal to the estimated coefficient divided by its standard error. The coefficients of distance, living area, and bedrooms all have t-values well above 2 and, therefore, two-sided $p < .05$.

c. No. Square footage is held constant but an extra bedroom could reduce the size of the living room, dining room, or other parts of the house without splitting an existing bedroom in half.

d. No. The coefficient of the number of bathrooms is *ceteris paribus*, holding living area constant.

15.14 The coefficients are *ceteris paribus*. The coefficient of X_4 is an estimate of the effect on home prices of a higher Walk Score, for a given square footage, number of bedrooms, and number of bathrooms.

15.15 The answers are in bold:

Variable	Mean	Median	Standard Deviation	Minimum	Maximum
Income	33,830	**24,246**	27,846	2,343	130,000
Sex	**.60**	0	.50	0	1
Race	.25	0	.72	0	**1**

N = **37,897.**

15.16 The researcher should identify the variables and show the value of each estimated coefficient.

15.17 a. A reasonable model is

$$Y = \alpha + \beta_1 D + \beta_2 X + \beta_3 D * X + \varepsilon$$

where:

    Y = total fine

    D = 1 if driver is under the age of 30, 0 otherwise

    X = miles over the speed limit

  b. Positive values of β_1 and β_3 would indicate discrimination against younger drivers.

  c. The average fine might be higher for younger drivers, but this is because they tend to drive more miles over the speed limit.

  d. The average fines are the same for younger and older drivers, even though older drivers tend to drive more miles over the speed limit.

15.18 a. The intercept α is the salary of a male with no teaching experience; β_1 is the difference between female and male salaries with no experience ($\alpha + \beta_1$ is the salary of a female with no teaching experience); β_2 is the increase in male salaries for each additional year of experience; β_3 is the difference between female and male salary increases for each additional year of experience ($\beta_2 + \beta_3$ is the increase in female salaries for each additional year of experience).

  b. A comparison of the group means doesn't take experience into account.

  c. Suppose salaries rise with experience. If there is no discrimination and males happen to have more experience than females, a regression equation can show no discrimination while a difference-of-means test does.

  d. If there is discrimination and males happen to have less experience than females, a regression equation can show discrimination while a difference-of-means test does not.

15.19 a. An R^2 close to zero suggests no relationship between X and Y, and that we should *not* reject the null hypotheses that the coefficients are zero. However, we can have an R^2 close to zero with t-values large enough to reject the null hypothesis. Here, the R^2 is not that close to zero and the t-values of 4.36 and 3.74 imply that we should reject the null hypotheses at the 5% level.

  b. In fact, the coefficient of X has the correct sign—a more favorable rating should lead to increased box office revenue. The coefficient does not seem unreasonably large: a rating of B+ instead of B is predicted to increase box office revenue by $6,453,431.

  c. The low t-value on the estimate of alpha means that the Y-intercept is not statistically significantly different from zero. But there is no reason why it should or should not equal zero. The estimate here implies that a G, PG, or PG-13 movie with an F rating from critics would have domestic box office revenue of $6,800,024. This doesn't seem unreasonable, and, even if it does, such a prediction probably involves incautious extrapolation, since very few movies are rated F.

15.20 If wealth belongs in the equation, then omitting it will bias the estimate of the coefficient of income and make the equation less useful for predicting spending.

15.21 Don't put in zeros for the missing values, as these will bias the estimates.

15.22 There should be separate 0-1 dummy variables for each state, instead of one dummy variable for all 50 states. There is no reason why D going from 1 to 2 (from Alabama to Alaska) has the same effect on household income as does D going from 2 to 3 (from Alaska to Arizona).

15.23 There is perfect multicollinearity since $M = 1 - U$. Either M or U should be dropped from the equation. It doesn't matter which one is dropped as long as researcher interprets the parameters carefully. Otherwise, the model cannot be estimated.

15.24 Theory should come before data. Choosing variables based on their statistical significance can, as here, lead to the inclusion of variables that are coincidentally correlated with the dependent variable and the omission of variables that truly belong in the model, but happen to not be statistically significant with the data used to estimate the model.

15.25 a. In general, multicollinearity among the explanatory variables increases the standard errors and reduces the t-values of the estimated coefficients.

   b. There cannot be a multicollinearity problem since there is only one explanatory variable.

15.26 They are all incorrect.

   a. We can't put a probability on the null hypothesis being true unless we do a Bayesian analysis. The p-value relates to the probability of observing these data if the null hypothesis is true, not the probability that the null hypothesis is true if we observe these data.

   b. A t-value gauges statistical significance, not whether the magnitude of the estimated coefficient is substantial.

   c. The R^2 can be low even if some of the explanatory variables are statistically significant.

   d. The category given the value $D = 1$ is arbitrary and does not affect the results?

15.27 a. $R^2 = .74$ means that 74% of the variation in heights is explained by the model.

b. The coefficient of X_1 shows that, for given parental heights, a male is predicted to be .434 feet taller than a female. The absolute value of the t-value merely tells us whether we can reject the null hypothesis that there is no difference between males and females, not the direction or size of the difference.

c. Yes. A student whose parents are each one-foot above average is predicted to be only .364 + .456 = .820 feet above average.

15.28 The low standard errors and high t-values (12.12/4.17 = 2.91 and 25.85/5.03 = 5.14) disprove the contention that multicollinearity renders the original estimates unreliable. When unemployment went up, interest rates also tended to go up and, together, these depressed stock prices by a large amount. It would be a mistake to omit interest rates from the estimated equation and attribute all of the decline in stock prices to the rise in unemployment. If the second equation is used to predict stock prices during a period when unemployment rises and interest rates don't, it will predict a much larger decline in stock prices than is likely to occur.

15.29 Removing a variable that should be included in an equation simply because it is correlated with another variable is likely to bias the estimated coefficients of the variable left in the model.

15.30 The omission of the rate of inflation will bias the estimated coefficient of the unemployment rate toward zero, reducing its absolute value. When unemployment increases, this tends to reduce the incumbent vote but also tends to reduce the rate of inflation, which increases the incumbent vote. These two effects on the incumbent vote tend to offset each other, so that it appears that an increase in the unemployment rate does not have as large a negative effect on the incumbent vote as it actually has.

15.31 Omitting important explanatory variables may bias the estimates of the coefficients of the included variables. The substantial drop in R^2 suggests that the omitted variables were important. Although the coefficient of female secondary education is now statistically significant, it should be troubling that it is likely to have the wrong sign. It is better to respond to a multicollinearity problem by adding more data than by dropping explanatory variables.

15.32 Despite the multicollinearity, the high t-values show that the explanatory variables have statistically persuasive effects on computer prices. (Whether these effects are substantial is a different question.)

15.33 Neither. The t-values indicate that there is stronger statistical evidence against the null hypothesis that home runs don't matter than against the null hypothesis that batting averages don't matter, but t-values don't gauge the size of the effects. The coefficient of batting average is larger than the coefficient of home runs, but the units are quite different. The coefficient of H indicates that one more home run is predicted to increase the salary by $.15 million, while the coefficient of batting average indicates that an increase in batting average by 1 (from .263 to 1.263, which is impossible) is predicted to increase salary by $13.24 million. We might, instead, use the means and standard deviations to calculate the elasticities at the mean values of the variables.

15.34 Housing starts trended downward by 1,300 per quarter, with strong seasonal patterns. Housing starts were higher by 90.93 thousand, 70.97 thousand, and 19.25 thousand in quarters 2, 3, and 4, respectively, relative to the first quarter (though the fourth quarter difference is not statistically significant at the 5% level). The estimated equation explains 19.6% of the variation in quarterly housing starts. The remainder is no doubt due to macroeconomic factors (like interest rates and the unemployment rate) that were not included in the model but have large effects on housing starts (including the big declines in the early 1980s and between 2005 and 2009).

15.35 a. With 100 − 6 = 94 degrees of freedom, the cutoff for a two-tail test of statistical significance at the 5% level is a t-value that is larger (in absolute value) than 1.99. Using the 1.99 cutoff, the coefficients of D, G, P, and S are statistically significant at the 5% level; the coefficient of A is not quite.

b. *A priori*, each of the five explanatory variables should have a positive effect on money winnings. Because all five estimated coefficients are positive, all five have plausible signs.

c. The intercept predicts the money winnings for a player who drives the ball zero yards, never hits the fairway, never reaches the green in par minus two, never requires only one putt, and never makes par out of the sand. Such a player would never be allowed on the PGA tour and it is an error of incautious extrapolation to predict the money winnings for such a terrible golfer. The intercept is set so that the predicted winnings of a player with average values of each of the five characteristics is equal to the average money winnings. With large positive values of each of the coefficients of the five explanatory variables, it turns out that a negative intercept is required.

d. Because the units of D and A are completely different, we cannot compare their effects simply by seeing which has the larger estimated coefficient. For example, if we were to measure D in feet rather than yards, the coefficient of D would be reduced by a factor of 3 but there would be no effect whatsoever on how well the model predicts money winnings.

e. The t-values gauge how confident we are that the coefficients are not equal to zero, not how much they affect winnings. Plus, as explained in (d), the units are completely different.

15.36 The coefficients are *ceteris paribus*. The coefficient of the dummy variable measures whether, for people with the same income, rent is affected by having a full-time job.

15.37 The coefficient of the dummy variable measures the extent to which extracurricular activity affects grades, holding study time constant (since study time is another explanatory variable in the equation). The coefficient of the dummy variable compares students who study equally, but one participates in an extracurricular activity in addition, and the other doesn't. (There might be an effect, for example, if the student is physically or emotionally drained by the extracurricular activity.)

15.38 a. The t-test is for the coefficient of Y.

   b. The value of β_2 is *ceteris paribus*, holding income constant.

   c. The two-sided p-value is close to .05.

   d. R^2 cannot be larger than 1.

   e. There is no single value of ε.

15.39 Regression estimates are not possible if all possibilities (like Frosh, Sophomore, Junior, and Senior) are included as dummy explanatory variables.

15.40 a. The explanatory variables should not be chosen on the basis of t-values.

   b. The magnitudes of the effects are gauged by the magnitudes of the estimated coefficients, not by R-squared or t-values.

   c. Unless we are Bayesians, we can't put probabilities on the null hypothesis being true.

   d. The multiple regression model does not assume that the explanatory variables are independent.

16

Miscellaneous

16.1 Identify the most important statistical problem with each of these studies; for example, non-response bias.

    a. The value of participating in college debate was demonstrated by comparing the GPAs of debaters with non-debaters.

    b. Correlations between a person's BMI and consumption of 100 different foods found that BMI was reduced by the consumption of apple juice.

    c. A 2021 study of the ten largest mutual funds found that they had outperformed the S&P 500 by more than two percentage points a year over the period 2000–2020.

    d. A 2020 cross-sectional study of the relationship between the average female unemployment rate and the daytime low temperature in 29 countries during the years 1997 through 2014 found a statistically significant negative correlation.

16.2 What is the most important statistical problem with each of these studies?

    a. An unusually large number of tweets of the word "happy" by Donald Trump tend to be followed by an increase in the level of the Dow Jones Industrial Average 3 days later.

    b. Heart disease fatalities were reported for the 8 heart-disease categories (out of 17 total categories) in which Asian-Americans have an above-average death rate on the fourth day of the month.

    c. A regression model found a statistically significant relationship between household spending and wealth after income was dropped as an explanatory variable.

    d. The benefits of studying French were demonstrated by the fact that high school students who had taken French classes scored 80 points higher, on average, than other students on the Reading/Writing SAT.

16.3 Here are several situations where there is an incorrect application of statistical reasoning. Explain briefly what is wrong and why it is wrong.

    a. A random sample of 100 students was asked if dining-hall food is better or worse this year than last year. A difference-in-proportions test compared the 57% who chose "improved" with the 43% who chose "worsened."

DOI: 10.1201/9781003630159-16

b. A one-sample t-test rejected the null hypothesis that the sample mean is 25.

c. A matched-pair t-test reported that the results are statistically significant at the 5% level because the p-value is .95.

d. A researcher calculated the price-earnings (P/E) ratios of the 30 stocks in the Dow Jones Industrial Average, as of January 1, 2017. A difference-in-means test was then used to compare the average price increase in 2017 for the 15 stocks with the lowest P/Es with the average price increase of all 30 stocks.

16.4 For which of the following tests do we need equal sample sizes?

a. Difference-in-means

b. Difference-in-proportions

c. Binomial

d. Chi-square

e. Regression

16.5 Do you agree or disagree with the following statements?

a. If R-squared in a simple regression model is positive, then the correlation R is positive.

b. In a chi-square test, if $\chi^2 = 1$, then $p = 0$.

c. P [A or B] is never equal to P [A] + P [B].

d. The variance of the sample mean of X cannot be larger than the variance of X.

e. The variance of a binomially distributed variable X cannot be larger than its mean.

16.6 Do you agree or disagree with the following statements?

a. If two sets of data have the same mean and standard deviation, their histograms are identical.

b. The median is never outside the interquartile range.

c. If A and B are independent, then P [A or B] = P [A] + P [B].

d. A normal distribution is symmetrical about its median.

e. The expected value of a probability distribution is the most likely outcome.

16.7 Do you agree or disagree with the following statements?

a. Since the p-value is .089, I accept the null hypothesis.

b. If X's mean is larger than Y's mean, then X's median is larger than Y's median.

c. In a box plot, an outlier is never inside the box.

    d. Multicollinearity is when a dependent variable is correlated with the explanatory variables.

    e. The explanatory variables in a multiple regression model have little effect on the dependent variable if R-squared is low.

16.8 Do you agree or disagree with the following statements?

    a. If R-squared is not high enough, you should add more explanatory variables.

    b. If there is a multicollinearity problem, you should drop some explanatory variables.

    c. If the null hypothesis is true, the p-value is more likely to be less than .10 than less than .05.

    d. The least squares line goes through the sample means of X and Y.

    e. The probability of A or B is equal to the probability of A plus the probability of B.

16.9 Explain why you either agree or disagree with each of these statements:

    a. A difference-in-means test assumes equal sample sizes.

    b. A difference-in-means test assumes equal standard deviations.

    c. A normal distribution is better than a binomial distribution for a single-sample test of a probability.

    d. A multiple regression model assumes that the explanatory variables are uncorrelated.

    e. A multiple regression model assumes that the explanatory variables are normally distributed.

16.10 Explain why you either agree or disagree with each of these statements:

    a. If I flip a fair coin until I obtain a head, on average, it will take me two flips.

    b. I used a normal approximation to the binomial distribution because the central limit theorem tells us that for large sample sizes, the distribution of the success probability p is approximately normal.

    c. If the expected value of the change in income is zero, then the change is equally likely to be positive or negative.

    d. The probability of 50 heads when a coin is flipped 100 times is larger than the probability of 5 heads when a coin is flipped 10 times.

    e. To investigate the relationship between interest rates and stock prices, I tested the null hypothesis that there is an effect against the alternative hypothesis that there is no effect.

16.11 Explain why you either agree or disagree with each of these statements:

    a. An ANOVA F-test and a difference-in-means t-test give the same p-value if there are two samples.

    b. Because the p-value is .01, I rejected the null hypothesis that the sample mean is –2.

    c. Because the p-value is larger than .05, we can reject the alternative hypothesis at the 5% level.

    d. The F-value will be negative if all the sample means are negative.

    e. I rejected the null hypothesis at the 5% level because the p-value is larger than .95.

16.12 Explain why you either agree or disagree with each of these statements:

    a. If I roll a six-sided die until I obtain a 6, it will, on average, take me six rolls.

    b. The average absolute deviation is less sensitive to outliers than is the standard deviation.

    c. The total area of the bars in a histogram is equal to 1.

    d. The central limit theorem assumes that X has a binomial distribution.

    e. I used a t-test instead of a Z-test because the sample size was so large.

16.13 Explain why you either agree or disagree with each of these statements:

    a. The central limit theorem explains why a binomial distribution with $p = .9$ converges to a normal distribution as n increases.

    b. For a normal distribution, there is a .95 probability of being within two standard deviations of the median.

    c. The width of the box in a box plot is equal to the interquartile range.

    d. The probability of four 2s when six dice are rolled can be determined by the binomial distribution.

    e. A t-value can never be negative.

16.14 Explain why you either agree or disagree with each of the following statements:

    a. A significance test that is significant at the 1% level is also significant at the 5% level.

    b. The ANOVA F-statistic can be used to test the null hypothesis that the sample means are equal.

    c. If the chi-square value is 0, then the p-value is 0.

    d. If the null hypothesis is p = .5, you can use a one-sided p-value once you know whether x/n is larger or smaller than .5.

    e. In a simple regression, R-squared is equal to the correlation coefficient squared.

16.15 Identify the most appropriate statistical test. You do not need to show any formulas, just identify the test, for example, "difference-in-means t-test":

    a. The average college student got less than 8 hours of sleep during the past 24 hours.

    b. The average price of Hollister clothing is less than the average price of essentially identical A&F clothing.

    c. People are more likely to have high blood pressure if they eat lots of dairy products and drink very little citrus juice.

    d. Taller male college students get better grades than do shorter male students.

    e. Horseshoe pitchers are more likely to throw a double-ringer if they threw a double-ringer in the previous inning than if they didn't throw a double-ringer.

16.16 Identify the most appropriate statistical test. You do not need to show any formulas, just identify the test, for example, "difference-in-means t-test":

    a. Daily trading volume tends to increase in December for stocks that have suffered large price declines between January and November.

    b. GPAs tend to be higher for the players on women's Division III college basketball teams than for the players on men's Division III college basketball teams.

    c. Holding earnings, risk, and growth rates constant, companies that pay higher dividends tend to have higher stock prices.

    d. The prices of the 30 stocks in the Dow Jones Industrial Average tend to increase between the close of trading each day and the opening of trading the next day.

    e. Japan's soccer goalie is more likely to dive to the left than to the right on penalty kicks.

16.17 Identify the most appropriate statistical test. You do not need to show any formulas, just identify the test, for example, "difference-in-means t-test":

    a. Professional poker players tend to play looser after they lose a big pot than after they win a big pot.

    b. The incumbent party's share of the presidential vote depends on the unemployment rate and the rate of inflation.

c. A company's stock price tends to go up on the Monday after it has an ad shown during the Super Bowl.

d. Baseball batting averages tend to increase over the course of a season.

e. Adult women are more likely to live in different ZIP codes from where they were born if their mother is a first-generation immigrant.

16.18 Identify the most appropriate statistical test. You do not need to show any formulas, just identify the test, for example, "difference-in-means t-test":

a. Time series data were used to predict interest rates based on the rate of inflation and the government deficit.

b. The daily returns on a professionally managed stock portfolio were compared to the daily returns on the S&P 500 index during the year 2000–2020.

c. "As January goes, so goes the year" is an old stock market adage. Data were collected for 1970–2020 on the S&P 500 return in January of each year and the S&P 500 return for February through December of that year.

d. In golf, each hole is assigned a par score (typically between 3 and 5) that a good golfer should be able to achieve. A study of several professional golf tournaments recorded how often a player's score on a hole was below-par, par, or above-par, and how often this player's score on the next hole was below-par, par, or above-par.

e. To test a mentalist's claim to be able to influence dice rolls, two standard six-sided dice were rolled 200 times and the number of doubles was recorded.

16.19 Identify the most appropriate statistical test. You do not need to show any formulas, just identify the test, for example, "difference-in-means t-test":

a. A study of 463 stock splits between January 1, 2007 and December 31, 2016, compared the return on these stocks on the day the split was announced to the return on the S&P 500 index that day.

b. A study predicted the number of hours a person spent on volunteer work each month based on the person's sex, race, age, and income.

c. A study compared average starting salaries for recent college graduates hired by Facebook, Amazon, Netflix, Google, and Apple.

d. The number of children a woman has (none, 1, 2, 3, 4 or more) is related to the number of children her mother had.

e. A study used data for all 58 California counties to compare the difference between the 2009 and 2017 divorce rate with the difference between the 2009 and 2017 unemployment rate.

16.20 Identify the most appropriate statistical test. You do not need to show any formulas, just identify the test, for example, "difference-in-means t-test":

a. Data for 48,440 adult patients infected with COVID-19 were used to compare the average age of those who died with the average age of those who did not die.

b. The infected patients were divided into three physical activity groups (consistently inactive, some activity, and consistently active) and the researchers compared the frequency with which people in each physical activity group were hospitalized.

c. The number of infected male and female patients were compared to see if infected patients are equally likely to be male or female.

d. The infected patients were divided into four age groups (< 60, 60–69, 70–79, and > 79) to see if there was a relationship between age and physical activity.

e. Each patient's probability of dying was predicted based on the patient's physical activity category and 25 other possible confounding variables.

16.21 Identify the most appropriate statistical test. You do not need to show any formulas, just identify the test, for example, "difference-in-means t-test":

a. A study of injuries suffered by varsity athletes during the years 1980–1995 obtained these data:

	Females	Males
Basketball	214	344
Soccer	155	236
Swimming	161	34
Tennis	82	82
Track	151	233

b. Data for 116 California communities were used to see how the percent change in home prices between 2005 and 2010 was affected by each county's percent change in home prices between 2000 and 2005 and each county's ratio of the median home price in 2005 to median rent in 2005.

c. A researcher looked at 50 stocks that had been removed from the Dow Jones Industrial Average and, in each case, identified whether, after the substitution, the stock that was removed did better or worse than the stock it replaced it.

d. Daily stock return data were used to see whether a portfolio of companies removed from the Dow Jones Industrial Average did better than a portfolio of the companies that replaced them.

e. Survey data were collected from 120 college students on the average time they went to bed each night during the previous semester and their GPA that semester.

16.22 Identify the most appropriate statistical test. You do not need to show any formulas, just identify the test, for example, "difference-in-means t-test":

a. More goals are scored in the second half of soccer games than in the first half.

b. A researcher investigated discrimination against women by a certain employer by comparing male and female salaries, taking into account education, years of experience, and other factors the employer claimed affected salaries.

c. Stocks with high price-earnings ratios at the peak of the dot-com bubble did worse than other stocks over the next 20 years.

d. A study compared male and female ratings (on a scale of 1–10) of Coke and Diet Coke.

e. A researcher investigated whether Jews are able to postpone their deaths until after Passover by counting the number of Jews who died during each of the 8 weeks surrounding Passover: 4 weeks before, 3 weeks before, 2 weeks before, 1 week before, 1 week after, 2 weeks after, 3 weeks after, and 4 weeks after.

16.23 Identify the most appropriate statistical test. You do not need to show any formulas, just identify the test, for example, "difference-in-means t-test":

a. A study compared the daily returns from a portfolio of the stocks recommended by the book, *Good to Great*, with the S&P 500 index.

b. A study investigated whether the fluctuations in the daily returns for ten stocks chosen by *Fortune* as the "most admired" could be explained by the overall stock market, each company's dividends, and each company's earnings.

c. A researcher investigated a strategy of betting against professional football teams that won the previous week by counting the number of bets won and lost using this strategy during the years 2000–2020.

d. A study compared the average income today of people who graduated 20 years ago from two different colleges.

e. A researcher investigated the hot hands phenomenon by tabulating how often professional bowlers followed a strike, spare, or open frame with a strike, spare, or open frame.

16.24 Identify the most appropriate statistical test. You do not need to show any formulas, just identify the test, for example, "difference-in-means t-test":

a. Oak trees are more likely to survive hurricanes than are pines trees or spruce trees.

b. The price of undeveloped land on the outskirts of a city depends on how close it is to the city center, and whether there are nearby water and electrical connections.

c. Identical brand-name products cost less at Costco than at Target.

d. The average age at death is the same for people with common and uncommon names.

e. A study compared daily stock-market returns on the New York Stock Exchange to the amount of cloudiness that day in New York (measured on a scale from 0 to 100).

16.25 Identify the most appropriate statistical test. You do not need to show any formulas, just identify the test, for example, "difference-in-means t-test":

a. Data collected by the Center for Coastal Studies found that female whales with an interval of less than 3 years between calving had 22 sons and 20 daughters. For whales calving at intervals of 3 or more years, there were 16 sons and only 4 daughters.

b. The ad revenue generated on a travel company's web page does not depend on whether the main page shows a picture of a beach, forest, or city.

c. A study of severe heart attacks suffered by German workers (either on or off the job) grouped the data according to the day of the week when the heart attack occurred.

d. The SAT scores and high school GPAs of 500 students were used to predict their first-year college GPAs.

e. In 1977, the US Supreme Court noted that persons with Spanish surnames constituted 79% of the population of Hidalgo County, Texas, but only 339 of 870 recent jurors in this county.

16.26 Identify the most appropriate statistical test. You do not need to show any formulas, just identify the test, for example, "difference-in-means t-test":

a. Math majors who play varsity sports have higher GPAs than do other math majors.

b. A baby's birth weight is related to the mother's pre-pregnancy weight.

c. A study compared the income of immigrant mothers and their daughters when each daughter was the same age as her mother at the time of the daughter's birth (e.g., if the daughter was born when her mother was 28, the study compared the mother and the daughter's income when each was 28).

d. The heights of Major League Baseball players do not depend on the position they play: catcher, pitcher, infield, or outfield.

e. The daily percentage change in the price of gold is independent of the percentage change the day before.

16.27 Identify the most appropriate statistical test. You do not need to show any formulas, just identify the test, for example, "difference-in-means t-test":

a. The sex of a mother's third child is independent of the sexes of her first two children.

b. An increase in the rate of inflation increases the interest rate on Treasury bills by the same amount.

c. People who smoke cigarettes are more likely to die of lung cancer than are nonsmokers.

d. A study tested the theory that two main determinants of stock prices are the unemployment rate and interest rates.

e. The daily returns from a portfolio of ten AI-powered mutual funds were compared to the S&P 500 from January 1, 2020, through December 31, 2024.

16.28 Identify the most appropriate statistical test. You do not need to show any formulas, just identify the test, for example, "difference-in-means t-test":

a. A black-box algorithm predicts whether the Dow Jones Industrial Average will go up or down each day.

b. A study compared the off-season and on-season GPAs of college students who play a varsity sports in either the spring or fall (but not both).

c. A study compared the success rates of soccer penalty shots taken to the left, right, and middle of the goal.

d. At-risk patients are less likely to have a fatal heart attack if they take aspirin daily instead of a placebo.

e. A study of the effect of quarries on home prices compared the prices of homes located at different distances from a quarry, taking into account each home's square footage, lot size, number of bedrooms, number of bathrooms, and the age of the home.

16.29 Identify the most appropriate statistical test. You do not need to show any formulas, just identify the test, for example, "difference-in-means t-test":

a. Home prices in Omaha depend on square footage, number of bedrooms, and number of bathrooms.

b. Thirty people label the pictures of Joe Biden shown on the front pages of the *New York Times* and the *New York Post* on the day after the 2020 presidential election as "favorable" or "unfavorable."

c. Essays written by 100 students in a college history class are graded on a scale of 1–100 by the professor and by an AI algorithm.

d. A researcher wanted to see if natural births are more likely at certain times of the day. She divided the day into 24 one-hour periods and recorded the number of births in each one-hour period.

e. A company's stock price usually rises after it announces the departure of its CEO.

16.30 Identify the most appropriate statistical test. You do not need to show any formulas, just identify the test, for example, "difference-in-means t-test":

a. Data were collected on the market values of 25 undamaged South Napa homes 6 months before and after the 2014 South Napa earthquake to see what effect, if any, the quake had on home values.

b. Daily stock returns in 2020 were collected for the S&P 500 index of stocks and for a portfolio of the publicly traded stocks identified by Glassdoor on December 10, 2019, as among the 50 best companies to work for.

c. It was estimated that .36% of the people living in Dublin have Brady as a last name and that 8 of the 999 people treated in Dublin hospitals for bradycardia (a slower than normal heart rate) have Brady as a last name.

d. Thirty-six subjects read a paragraph about an elderly man named George. For 18 people, the paragraph described him as competent; for the other 18, he was described as incompetent. After reading the story, each subject indicated how warm and friendly George is on a scale of 1 (not at all) to 9 (very).

e. After its development, the Pfizer mRNA vaccine was tested through a randomized controlled trial of 37,586 participants, with 18,801 people in the treatment group given the vaccine and 18,785 people in the control group given an injection of a saline placebo. There were 8 cases of COVID-19 in the treatment group, compared to 162 in the control group.

16.31 Identify the most appropriate statistical test. You do not need to show any formulas, just identify the test, for example, "difference-in-means t-test":

  a. There are more motorcycle accidents on Friday nights with a full moon than on Friday nights with no moon.

  b. A researcher looked at daily returns in the stock market over the past 20 years to see if returns tend to be relatively good or bad on certain days of the week.

  c. A researcher looked at the performance of stocks with clever ticker symbols (such as MOO for United Stockyards) by comparing the daily returns for a portfolio of clever ticker stocks with the S&P 500.

  d. A survey of 100 college students compared the percentage who were first-born or only children to the national percentage of 40%.

  e. A survey of 150 college students compared the number of first-generation, second-generation, and other college attendees with their choice of a STEM major or a non-STEM major.

16.32 Identify the most appropriate statistical test. You do not need to show any formulas, just identify the test, for example, "difference-in-means t-test":

  a. A researcher investigated the socioeconomic mobility of California women by using data on each woman's income, her parent's income, whether or not her mother was born in the United States, and whether or not her father was born in the United States.

  b. Twenty-four students were asked to read a short article and when they were done, were asked to guess how long it took them to read the article; 21 of 24 students overestimated the time.

  c. The time it took each of the 24 students in part (b) to read the article was compared to his or her guess.

  d. An Internet consulting company tested commercial web page layouts by randomly directing users to different pages; for example, a background that is a light blue, light red, light green, or white. The company then measured the click-through rate for each layout (the fraction of users who click on a link that gives them more information about a product).

  e. A researcher compared the price/earnings (P/E) ratios of 50 popular ("Nifty 50") stocks on December 31, 1972, and the percentage returns on each of these 50 stocks from 1973 through 2022.

16.33 A data set has 47 observations, ranging from 2 to 658 with a median of 99. If the 2 were changed to 12, would each of the following increase, decrease, or stay the same?

a. Mean

b. Median

c. Standard deviation

d. Interquartile range

16.34 A study asked 60 students to predict their scores on a high school math test. It turned out that:

a. The average actual score (59.1) was lower than the average pre-dicted score (74.9).

b. There was a .65 correlation between the predicted and actual scores.

The author concluded that

a. Students over-estimate their ability.

b. Test scores can be increased by raising students' self-esteem:

> *The educational implication of this observation is that there is correlation between self-esteem and self-efficacy, suggesting that teachers strive to raise students' self-esteem in mathematics by way of motivation.... This observation was confirmed by the fact that two students who predicted that they would fail actually did so. This implies that there is also a cor-relation between negative self-perception or inferiority complex and inad-equate motivation to learn or failure.*

Provide completely different statistical interpretations of results (a) and (b).

16.35 In 1996 The Motley Fool advised readers that their Foolish-Four sys-tem, based on data for 1973 to 1995, "should grant its fans the same 25% annualized returns going forward that it has served up in the past." The Foolish Four system is:

> On January 1, calculate the dividend yield for each Dow stock.

> Choose the ten stocks with the highest dividend yields.

> Of these ten, choose the five with the lowest price per share.

> Of these five, cross out the one with the lowest price.

> Invest 40% in the stock with the next lowest price.

> Invest 20% of your wealth in each of the other three stocks.

Why, as a statistician, are you skeptical?

16.36 In August 2015, *Sports Illustrated* reported that 2015 had been an incred-ible sports year that made "a mockery of the odds." They looked at the five remarkable events below and wondered: "What are the chances of these things happening individually? What about in the same year? Statistician Ed Feng quantified the likelihood—or lack thereof" based on the betting odds before each event:

US women's soccer team winning World Cup: 1 in 35

American Pharaoh winning Triple Crown: 1 in 9.9

Kentucky men's college basketball team going 38-0: 1 in 17.7

Serena Williams winning tennis' grand slam: 1 in 70.4

Jordan Spieth winning 2 of 4 majors: 1 in 3,118.5

Feng calculated the probability of all five events happening to be 1 in 522,675. As a statistician, are you convinced that 2015 made a mockery of the odds?

16.37 A researcher wrote that, "It remains apparent that the model needs to be manipulated and tweaked. As researchers, the next step in our study is to obtain statistically significant results." Why is this bad advice?

16.38 If a company is considering changing the main color on a web page from, say, blue to red, green, or purple, in order to increase the number of users who "click through" to another page, BestWeb sets up four pages using these colors. When a user goes to the page, a random number generator is used to take the user to one of the four pages and BestWeb records the click-through rate (fraction of users who click through). If there is a statistically significant difference, the client is advised to use the color with the highest click-through rate.

a. What is the appropriate statistical test?

b. When a client follows BestWeb's advice to change the predominant color of its web page, the client often finds that the actual increase in the click-through rate is not as large as observed in BestWeb's experiment. How would a statistician explain this?

16.39 In *Good to Great*, Jim Collins and his research team spent 5 years looking at the 40-year history of 1,435 companies and identified 11 stocks that outperformed the overall market. Collins then identified five distinguishing traits, such as Level 5 Leadership (leaders who are personally humble, but professionally driven to make a company great). He wrote that

> *we developed all of the concepts in this book by making empirical deductions directly from the data. We did not begin this project with a theory to test or prove. We sought to build a theory from the ground up, derived directly from the evidence.*

To buttress the statistical legitimacy of his research, Collins talked to two professors at the University of Colorado. One said that, "the probabilities that the concepts in your framework appear by random chance are essentially zero." The other professor was more specific: "What is the probability of finding by chance a group of 11 companies, all of whose members display the primary traits you discovered while the direct comparisons do not possess those traits?" He calculated this probability to be less than 1 in 17 million.

What is the logical error in this calculation? Do not check the math; identify the logical error.

16.40 John Gottman has written several books, given many talks, and, with his wife, created The Gottman Institute for marriage consulting and therapist training. In a 2007 survey of psychotherapists, Gottman was voted one of the ten most influential members of their profession over the past 25 years. In his seminal study, 130 newly-wed couples were videotaped while they had a 15-minute discussion of a contentious topic. Gottman went over the videotapes, frame by frame, recording detailed changes in facial expressions, tone of voice, and the like—for example, noting whether the corners of a person's mouth were upturned or downturned during a smile. Six years later, he identified the couples that had divorced and created a model for predicting divorce based on the codings he had made 6 years earlier. He reported that his model was 82.5% accurate. Malcolm Gladwell said that, "He's gotten so good at thin-slicing marriages that he says he can be at a restaurant and eavesdrop on the couple one table over and get a pretty good sense of whether they need to start thinking about hiring lawyers and dividing up custody of the children."

Why should we be skeptical of Gottman's procedure?

16.41 A finance PhD student analyzed monthly data during the years 1995 to 2014 for two home price indexes, one ("P") for homes in the prime London neighborhoods of Kensington and Chelsea, and the City of Westminster and the other ("L") for homes in Liverpool. The home price indexes were both scaled to equal 100 in January 1995. Her descriptive statistics were

	Liverpool Index L	Prime Index P
Mean	145.7	339.0
Standard deviation	26.6	138.9
Minimum	100.0	100.0
Maximum	191.4	578.0

She also calculated the luxury ratio each year

$$\text{Luxury ratio} = 100\frac{P}{L}$$

and constructed this graph:

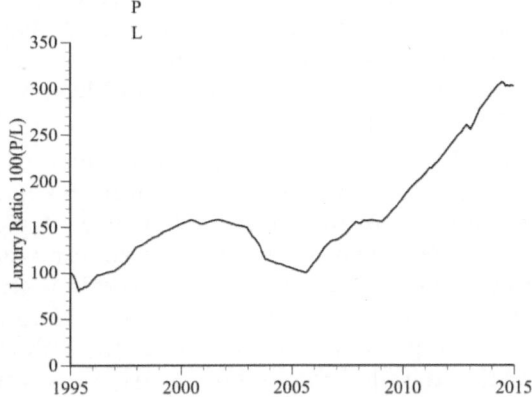

Critically evaluate each of these conclusions:

a. "The standard deviation is larger for prime properties because of the greater diversity of more expensive homes in prime London neighborhoods as compared with the Liverpool housing market."

b. "The luxury ratio of prime London with Liverpool represents the average excess amount, in percentage terms, that is paid in prime London relative to Liverpool for a property. A luxury ratio of 150, for example, signifies that the price level of prime property is 50% greater than the price in Liverpool."

c. "We can also observe that 2005 was the right moment to buy prime property in London if one meant to do so. Premiums paid in prime London with respect to Liverpool greatly increased after 2005."

16.42 A statistician looked at mortality data for a large city in Ohio and identified the home addresses of people who had died of cancer within the past 50 years, represented by dots in this map:

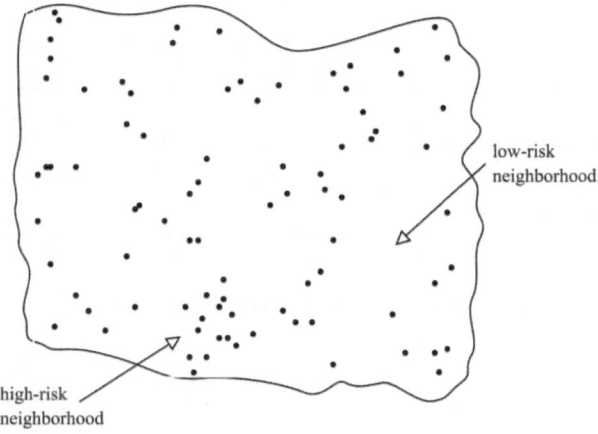

After identifying a high-risk neighborhood and a low-risk neighborhood, she found that there were many more live-oak trees in the high-risk neighborhood. She did a difference-in-proportions test comparing the cancer frequencies in these two neighborhoods and got a Z-value of 3.89 and a two-sided p-value of .0001. A journal published her results, including her recommendation that cities reduce cancer risks by removing live-oak trees. How would you respond? Make a clear argument that might persuade a city council.

16.43 A study reported that among Major League Baseball players born between 1875 and 1930, players whose first names began with the letter D died, on average, 1.7 years younger than did players whose first names began with the letters E through Z. What statistical reasons make you skeptical?

16.44 A Yale economics professor and graduate student looked at Bitcoin prices from January 1, 2011 to May 31, 2018 (as far back as there were reliable data) and calculated correlations between Bitcoin returns and 810 other financial variables. They found that 63 had p-values below .10, including a positive correlation between bitcoin returns and stock returns in the consumer goods and health care industries and a negative correlation with stocks returns in the fabricated products and metal mining industries. Why are you skeptical?

16.45 A study of the relationship between socioeconomic status and juvenile delinquency tested 756 possible relationships and found 33 to be statistically significant at the 5% level. What statistical reason is there for caution here?

16.46 Suppose that 1% of the null hypotheses tested by a researcher are false and 99% are true. Assume that the outcomes of the tests are independent and that, for each test, there is a .05 probability of rejecting the null hypothesis if it is true and a .05 probability of not rejecting a null hypothesis if it is false.

a. What fraction of the null hypotheses tested by this researcher will be rejected?

b. What fraction of all the null hypotheses that this researcher rejects are in fact true?

16.47 Paul the Octopus went 8 for 8 in predicting the winner of soccer games at the 2010 World Cup. Assuming independence, use these data to test the null hypothesis that Paul has a .5 probability of picking the winner of a soccer game. Why might p-hacking and HARKing be problems here?

16.48 Smith has 47 songs on his Apple Music playlist. Five are Tom Petty songs. He has been told that the Shuffle command puts the 47 songs in random order and will play all 47 songs before repeating any song. If this is truly how the Shuffle command works, and if he turns on Apple Music right now and chooses Shuffle, what is the probability that the first three songs he hears will be Tom Petty songs? If he does hear three Tom Petty songs in a row, is this statistically persuasive evidence that the Shuffle command is not as described?

Answers

16.1 a. Self-selection bias

   b. HARKing

   c. Survivorship bias

   d. P-hacking

16.2 a. HARKing

   b. P-hacking

   c. Omitted-variable bias

   d. Self-selection bias

16.3 a. It should be a one-sample test of the null hypothesis that students are equally like to choose "improved" and "worsened."

   b. Statistical tests are used for a null hypothesis about the population mean, not the sample mean.

   c. For statistical significance at the 5% level, the p-value should be less than .05.

   d. These are not independent samples since the 15 stocks are part of the 30; instead, compare the 15 stocks with the lowest P/Es to the 15 stocks with the highest P/Es.

16.4 We need equal sample sizes only for regression, where we must pair the X and Y values.

16.5 a. No, the two variables can be negatively correlated.

   b. No, if $\chi^2 = 0$, then $p = 1$.

   c. No, P [A and B] = 0 if A and B are mutually exclusive.

   d. Yes, variance of $\bar{X} = \dfrac{\text{variance of X}}{n}$.

   e. Yes, variance of $X = p(1-p)n$; mean of $X = pn$.

16.6 a. No.

   b. Yes.

   c. No (e.g., with two independent coin flips, P [H1 or H2] $\neq$ P [H1] + P [H2]).

d. Yes (A normal distribution is symmetrical and the median equals the mean).

e. No. The expected value is the long-run average value.

16.7 a. No. We don't accept a null hypothesis; we fail to reject.

b. No. For example, X might have a lower median but a larger outlier.

c. Yes. An outlier must be far outside the box.

d. No. Multicollinearity is when some of the explanatory variables are highly correlated with each other.

e. No. The size of the effects depends on the size of the coefficients, not the size of R^2.

16.8 a. No. Variables should be chosen for logical reasons and overfitting should be avoided.

b. No. This is likely to bias the estimates of the coefficients of the remaining explanatory variables.

c. Yes.

d. Yes.

e. No. We have to subtract P [A and B].

16.9 They are all false.

16.10 a. Yes. The expected wait is $1/p = 2$ for $p = .5$.

b. No. The success probability is a fixed parameter; it does not have a distribution.

c. No. For example, a .8 probability of a one-point increase and a .2 probability of a four-point decline has an expected value of zero.

d. No. As the number of flips increases, the probability of exactly 50% heads declines.

e. No. The appropriate null hypothesis is that there is no effect. The problem with having the null hypothesis be that there is an effect is that there are so many possibilities for the size of the effect.

16.11 a. Yes. $F = t^2$ and the p-value for the F-test is equal to the two-sided p-value for the t-test.

b. No. The null hypothesis concerns the population mean, not the sample mean.

c. No. We reject the null hypothesis at the 5% level if the p-value is less than .05. If the p-value is larger than .05, then we do not reject the null hypothesis, but this is not the same as rejecting the alternative hypothesis.

d. No. An F-value can never be negative.

e. No. You reject the null hypothesis at the 5% level if the p-value is less than .05.

16.12 a. Yes. The expected wait is $1/p = 6$ for $p = 1/6$.

   b. Yes. The squaring of deviations makes the standard deviation more sensitive to outliers.

   c. Yes.

   d. No. It applies to any distribution (with finite variance).

   e. No. The t-test is used when the population standard deviation is estimated and is more appropriate than a Z-test if the sample size is small. For large samples, they are identical.

16.13 a. Yes. The binomial distribution converges to a normal distribution as n increases for all values of p.

   b. Yes. There is a .95 probability of being within two standard deviations of the mean and, with a normal distribution, the mean is equal to the median.

   c. Yes. The interquartile range is the difference between the first and third quartiles and these are the edges of the box.

   d. Yes. These are Bernoulli trials with $n = 6$, $x = 2$, and $p = 1/6$.

   e. No. A t-value can be negative, for example, in a one-sample test where the sample mean is less than the null-hypothesis value of the population mean.

16.14 a. Yes. If the p-value is less than .01, then it is also less than .05.

   b. No. The ANOVA F-statistic tests the null hypothesis that the population means are equal.

   c. No. If the chi-square value is 0, then the p-value is 1, since this provides no evidence whatsoever against the null hypothesis.

   d. No. You should decide whether to use a one-sided or two-sided p-value before you look at your data.

   e. Yes.

16.15 a. One-sample t-test of $\mu = 8$.

   b. Matched-pair one-sample t-test of $\mu = 0$.

   c. Multiple regression.

   d. Simple regression.

   e. Difference-in-proportions Z-test or chi-square test.

16.16 a. Simple regression.

   b. Two-sample difference-in-means t-test.

   c. Multiple regression.

   d. Matched-pair 1-sample t-test of $\mu = 0$.

   e. One-sample binomial test of $p = .5$.

16.17 a. Difference-in-means t-test.
    b. Multiple regression.
    c. One-sample t-test.
    d. Simple regression.
    e. Difference-in-proportions Z-test or chi-square test.

16.18 a. Multiple regression.
    b. Matched-pair 1-sample t-test of $\mu = 0$.
    c. Simple regression.
    d. Chi-square test.
    e. One-sample binomial test of $p = 1/6$.

16.19 a. Simple regression.
    b. Multiple regression.
    c. ANOVA or multiple regression with four dummy variables.
    d. Chi-square.
    e. Simple regression.

16.20 a. Two-sample t-test.
    b. Chi-square.
    c. One-sample binomial test.
    d. Chi-square.
    e. Multiple regression.

16.21 a. Chi-square.
    b. Multiple regression.
    c. One-sample binomial test of $p = .5$.
    d. Matched-pair 1-sample t-test of $\mu = 0$.
    e. Simple regression.

16.22 a. Matched-pair 1-sample t-test of $\mu = 0$.
    b. Multiple regression.
    c. Simple regression.
    d. Difference-in-means t-test.
    e. Chi-square.

16.23 a. Matched-pair 1-sample t-test of $\mu = 0$.
    b. Multiple regression.
    c. One-sample binomial test of $p = .5$.
    d. Difference-in-means t-test.
    e. Chi-square test.

16.24 a. Chi-square.

      b. Multiple regression.

      c. Matched-pair 1-sample t-test of $\mu = 0$.

      d. Difference-in-means t-test.

      e. Simple regression.

16.25 a. Difference-in-proportions Z-test or chi-square.

      b. ANOVA or multiple regression with dummy variables.

      c. Chi-square.

      d. Simple regression.

      e. One-sample binomial test of $p = .79$.

16.26 a. ANOVA or difference-in-means t-test.

      b. Simple regression.

      c. Matched-pair 1-sample t-test of $\mu = 0$.

      d. ANOVA or multiple regression with dummy variables.

      e. Simple regression.

16.27 a. Chi-square.

      b. Simple regression.

      c. Difference-in-proportions Z-test or chi-square test.

      d. Multiple regression.

      e. Matched-pair 1-sample t-test of $\mu = 0$.

16.28 a. One-sample binomial test of $p = .5$.

      b. Matched-pair 1-sample t-test of $\mu = 0$.

      c. Chi-square.

      d. Difference-in-proportions Z-test or chi-square.

      e. Multiple regression.

16.29 a. Multiple regression.

      b. Difference-in-proportions Z-test or chi-square.

      c. Simple regression.

      d. Chi-square.

      e. One-sample binomial test of $p = .5$.

16.30 a. Matched-pair 1-sample t-test of $\mu = 0$.

      b. Matched-pair 1-sample t-test of $\mu = 0$.

      c. One-sample binomial test of $p = .36$.

      d. Difference-in-means t-test.

      e. Difference-in-proportions Z-test or chi-square.

16.31 a. Difference-in-means t-test.

b. ANOVA or multiple regression with dummy variables.

c. Matched-pair 1-sample t-test of $\mu = 0$.

d. One-sample binomial test of $p = .4$.

e. Chi-square.

16.32 a. Multiple regression.

b. One-sample binomial test of $p = .5$.

c. Matched-pair 1-sample t-test of $\mu = 0$.

d. Chi-square.

e. Simple regression.

16.33 a. Increase.

b. No change.

c. Decrease.

d. No change.

16.34 a. An alternative explanation is that the students underestimated the difficulty of the test that they would be given. It would have been more interesting to ask students to estimate their percentile score relative to the other students in the class. If 70% think they are above average, this would be convincing evidence that they overestimate their ability.

b. An alternative explanation is that students know something about their mathematical abilities. Most who did well know they are good at math. The students who failed knew that they did not know the material well enough to pass. A valid test might divide the students randomly into two groups, one of which is given positive reinforcement, while the other is given none (or criticized for their inadequacies).

16.35 The complexity of this system made it appear scientific. In reality, the complexity was needed to make the coincidental correlation stronger—like counting the Pittsburgh Steelers as an NFC team in the Super Bowl Indicator. This is clearly HARKing, choosing a strategy that fits the data, in that no one would have concocted this complicated system before looking at the data.

(Two finance professors looked at data for 1949–1972. The Foolish Four was a flop. They looked at stocks chosen on July 1 instead of January 1. The Foolish Four was again a flop. In 1997, the Gardners replaced the Foolish Four with the UV4. In 2000, the Gardners buried the Foolish Four and the UV4.)

16.36 It is not clear how they came up with 522,675, since the product of these five probabilities is 1/1,346,464,516. Nonetheless, the more fundamental problem is that the events were chosen after they happened. If *Sports Illustrated* (or Ed Feng) had predicted these five events before they happened, that would have been incredible. Not so, predicting them after they already occurred. There were thousands of sporting events in that year and the probability of being able to identify some unlikely events after they occurred is one.

16.37 The model should generally come before the data. It is misleading to test a model with the data used to derive the model.

16.38 a.   Chi-square test.

b.   Regression toward the mean.

16.39 This probability calculation assumes that the traits were specified before looking at the data, not afterward. The probability that there will be some common traits is 1.

16.40 This is HARKing. Gottman didn't actually predict whether a couple would get divorced. His models "predicted" whether a couple had already gotten divorced—which is a lot easier when you already know the answer. Gottman data-mined his detailed codings, looking for the variables that were the most highly correlated with divorces that had already happened.

16.41 a.   The table shows the standard deviation of each index across time, not the standard deviation of the prices of the houses in the index.

b.   A luxury ratio of 150 means that prime prices have increased 50% more than non-prime prices, and tells us nothing about the cost of buying prime and non-prime properties. This is likely saying that an index of car prices is 200 and an index of apple prices is 400; therefore an apple costs twice as much as a car.

c.   The price of prime property relative to non-prime property could have increased even while the price of prime property is falling. Suppose prime is $1 million and non-prime is $500,000 initially, and prime falls to $500,000 while non-prime falls to $200,000. The ratio has increased from 2.0 to 2.5 even while prime was a disastrous investment (though not as disastrous as nonprime). Buying at $1,000,000 was not "the right moment to buy prime property in London if one meant to do so." In addition, the return from housing depends on the rent saving, property taxes, etc.

16.42 Coincidental clusters can be found even in random data.

16.43 The data seem to have been tortured by omitting data that did not support this silly theory. Why the letter D? Why 1875–1930? Why first names?

16.44 This is clearly HARKing. The 63 correlations with p-values below .10 are seldom sensible and are fewer than the 81 that would be expected if they just correlated bitcoin prices with random numbers.

16.45 This is just HARKing. If the 756 possible relationships were all untrue and independent, we expect .05(756) = 37.8 to be statistically significant at the 5% level.

16.46 In general, suppose that a fraction λ of the null hypotheses tested by a researcher are false, there is a probability α of rejecting the null hypothesis if it is true, and there is a probability β of not rejecting a null hypothesis if it is false. In this exercise, $\lambda = .01$, $\alpha = .05$, and $\beta = .05$.

  a. The probability of rejecting a null hypothesis is
P [reject H_0] = P [H_0 is true] P [reject H_0 if H_0 is true] + P [H_0 is false] P [reject H_0 if H_0 is false] = $(1 - \lambda) \alpha + \lambda (1 - \beta)$. For $\lambda = .01$, $\alpha = .05$, and $\beta = .05$, this probability is .059.

  b. Using Bayes' rule,

P[H_0 true if reject H_0]

$$= \frac{P[H_0 \text{ true}]P[\text{reject } H_0 \text{ if } H_0 \text{ true}]}{p[H_0 \text{ true}]P[\text{reject } H_0 \text{ if } H_0 \text{ true}] + P[H_0 \text{ false}]P[\text{reject } H_0 \text{ if } H_0 \text{ false}]}$$

$$= \frac{(1 - \lambda)\alpha}{(1 - \lambda)\alpha + \lambda(1 - \beta)}$$

For $\lambda = .01$, $\alpha = .05$, and $\beta = .05$, this probability is .839.
Using a contingency table:

	Reject H_0	Do Not Reject H_0	Total
H_0 true	495	9,495	9,900
H_0 false	95	5	100
Total	590	9,500	10,000

  a. P [reject H_0] = 590/10,000 = .059.

  b. P [H_0 true if reject H_0] = 495/590 = .839.

16.47 If the null hypothesis is true, the probability of 8 out of 8 correct predictions is $.5^8 = .004$. The two-sided p-value is 2 (.004) = .008. There is undoubtedly p-hacking and HARKing in that Paul may have predicted many other games, perhaps in other sports, that were not reported because he was not so successful. In addition, many animals may have been tested by many people, with only the lucky successes reported. In the 2010 World Cup, there were news stories about several psychic animals, including Mani the Parakeet, Pino the Chimp, Apfelsine the Red River Hog, and Lin Ping the Psychic Panda.

16.48 The probability of three Tom Petty songs in a row is (5/47) (4/46) (3/45) = .0006167. This probability is low but it is not statistically persuasive because the chances that Smith will hear three songs in a row by some artist at some point is much larger than .0006167. (Suppose, for example, that his playlist contains 5 Tom Petty songs and 42 Bruce Springsteen songs.)

17

Out-of-Class Projects

Mean, Median, Descriptive Statistics, and Graphs

1. Morningstar and other Internet web sites have data on the total returns (dividends plus capital gains) for mutual funds in various fund categories (e.g., US Equity Small Value). Pick a category within US Equity Funds and select a horizon (e.g., 5 years). Summarize your data with a box plot and histogram, and compare the mean and median returns for the funds in your sample to the return for the S&P 500 index that year. (Make sure that your S&P 500 return is not just the change in the S&P, but the S&P total return, dividends plus capital gains.) Also calculate the fractions of your sample that are within one and two standard deviations of the mean return for your data.

2. Go to a local grocery store and collect these data for at least 50 breakfast cereals: cereal name; grams of sugar per serving; and shelf location (bottom, middle, or top). Group the data by shelf location and use three box plots to compare the sugar content by shelf location.

3. At a local grocery store, identify two categories of chips (e.g., potato chips and tortilla chips) and, in each category, find the grams of fat and the milligrams of sodium per serving on the nutritional labels of at least ten different varieties. (Do not include any low-fat or low-salt brands.) Use two box plots to compare the fat in these two categories; do the same with the sodium data.

4. Go to a local grocery store and collect data for at least 20 different soups from each of two major soup makers; for example, Campbell's and Progresso. For each of these soups, record the per-serving amounts of calories, fat, and sodium. Summarize these data.

5. Display some interesting data from the most recent *US News & World Report* college rankings (or a similar source); for example, the distribution of resources/spending per student.

6. Summarize and display some cross-section or time series data on deaths due to COVID-19.

7. The real interest rate is (approximately) equal to the nominal (dollar) interest rate minus the rate of inflation. Some people argue that inflation and the interest rates on Treasury bonds move up and down together so that real interest rates are constant. Summarize and

DOI: 10.1201/9781003630159-17

display some data that might shed light on this assertion. (One data source is https://fred.stlouisfed.org/.)

8. Some people argue that the real inflation-adjusted interest rates on Treasury bonds are constant. The returns from Treasury Inflation-Protected Securities, or TIPS, are adjusted for inflation so that the quoted interest rates are real interest rates. Summarize and display some TIPS data that might indicate whether real interest rates are constant. (One data source is https://fred.stlouisfed.org/.)

9. Summarize and display data on the calories, fat, and salt equivalent in a McDonald's Big Mac in at least 20 countries, such as the United States, United Kingdom, Canada, Chile, Japan, Israel, and Turkey.

10. Summarize and display data on the seasonally adjusted unemployment rate in the United Kingdom since 1855. (One data source is https://fred.stlouisfed.org/.)

11. Summarize and display data on consol (long-term bond) yields in the United Kingdom since 1753. (One data source is https://fred.stlouisfed.org/.)

12. Summarize and display data on the monthly percentage changes in the Share Price Index (weighted by market capitalization) in the United Kingdom since 1709. (One data source is https://fred.stlouisfed.org/.)

13. Compare the monthly percentage changes since 1914 in the Share Price Index (weighted by market capitalization) in the United Kingdom and the Dow-Jones Industrial Stock Price Index for the United States. (One data source is https://fred.stlouisfed.org/.)

14. Compare the monthly changes in the seasonally adjusted Case-Shiller NV-Las Vegas, TX-Dallas, and FL-Miami Home Price Indexes. (One data source is https://fred.stlouisfed.org/.)

15. Compare the monthly changes in the seasonally adjusted Case-Shiller CA-Los Angeles, CA-San Diego, and CA-San Francisco Home Price Indexes. (One data source is https://fred.stlouisfed.org/.)

16. Compare the monthly changes in the seasonally adjusted Case-Shiller CA-San Francisco, MA-Boston, and NY-New York Home Price Indexes. (One data source is https://fred.stlouisfed.org/.)

Probabilities

1. Flip a coin 200 times (use a real coin, not a computer simulation). Divide your results into 20 groups of 10 flips, and count the number of heads in each of these 20 groups. How often did you get ten heads

in ten flips? Nine heads? And so on. Use a histogram to summarize your results. Now identify the longest streak of consecutive heads or consecutive tails in each of these 20 groups. How often did you get a streak of ten? A streak of nine? And so on. Use a histogram to summarize your results.

2. Baseball-reference.com is one source of detailed historical statistics for Major League Baseball (MLB) players. Select a completed season and identify the players who had 502 times at bat that season (the minimum required to qualify for season batting awards). Find the batting averages and wins above replacement (WAR) for each of these players. Display your data in histograms and calculate the means and standard deviations. What fractions of the batting averages and WARs are within one, two, and three standard deviations of the mean?

3. Baseball-reference.com is one source of detailed historical statistics for MLB players. Select a completed season and identify the players who pitched at least 162 innings that season (the minimum required to qualify for season pitching awards). Find the earned run averages (ERA) and WAR for each of these players. Display your data in histograms and calculate the means and standard deviations. What fractions of the ERAs and WARs are within one, two, and three standard deviations of the mean?

4. Baseball-reference.com is one source of detailed historical statistics for MLB players. One way to compare outstanding players from different historical periods is to see how many standard deviations each player's performance was from the average performance that year. Considering only the batting averages and WAR of players who had at least 502 times at bat that season (the minimum number required to qualify for season batting awards), compare Miguel Cabrera's performance in 2015, Tony Gwynn's performance in 1996, and Ted Williams' performance in 1941.

5. Baseball-reference.com is one source of detailed historical statistics for MLB players. One way to compare outstanding players from different historical periods is to calculate how many standard deviations each player's performance was from the average performance that year. Considering only the ERA and WAR of players who pitched at least 162 innings that season (the minimum required to qualify for season pitching awards), compare Zack Greinke's performance in 2015, Greg Maddux's performance in 1995, and Sandy Koufax's performance in 1966.

6. Identify the two players in the American League and the two players in the National League who hit the most home runs in the most recently completed MLB season. For each of these players, use the data for the three most recent seasons to estimate the probability

p that this player will hit a home run during a time at bat. Also use your data for each of these four players over all three seasons to estimate the average number of times at bat n during a season. Now, assuming the binomial model is appropriate, use your estimates of p and n to estimate each player's chances of hitting more than Aaron Judge's (non-steroid-assisted) 2022 record of 62 home runs in a season.

7. Stock returns are the *product* of returns over shorter horizons. The logarithm of returns is the *sum* of the logarithm of returns over shorter horizons and may be governed by the normal distribution. Roll two standard six-sided dice 200 times and record the product of the numbers on the two dice. Now calculate the natural logarithm of each of these 200 products and display your results in a histogram.

8. You offer a friend a choice of three bags—one containing a candy bar or some other desirable prize and the other two bags containing already-been-chewed (ABC) gum. You know which bag contains the good prize, but your friend does not. After your friend picks a bag (and before looking inside), show your friend a bag that was not picked and contains ABC gum and ask your friend if they want to change their choice. Play this game with a large number of friends and record: (a) how often people switch; and (b) how often they would have won if they had switched every time.

9. Mr. Smith is walking with his daughter and has a second child at home. It has been argued that the probability that the child at home is a girl is either one-third or one-half. What do you think? Gather your own data by asking a large number of people how many siblings they have. If they have one sibling, record the sex of the person and the sibling. Is the frequency with which the sexes are the same closer to one-third or one-half?

10. You have three cards: one black on both sides, one white on both sides, and one white on one side and black on the other side. After the three cards are shuffled and dropped into an empty bag, you pull one card out and look at one side of the card. What is the probability that the back of the card is the same color as the front of the card? Play this game a large number of times and see if your results confirm or contradict your expectations.

One-Sample Confidence Intervals

1. Estimate the percentage of the seniors at your school who regularly follow the daily news, the percentage who can name the two US senators from their home state, the percentage who are registered to

vote, and the percentage who would almost certainly vote if a presidential election were held today.

2. Among seniors at your school who are looking for jobs, estimate the average annual salary they expect to earn their first year. Do not include moving allowances or other one-time benefits.

3. A study by Camilla Benbow and Julian Stanley of children younger than 14 who scored 700 or higher on the math SAT found that 20% were left-handed, as compared to 8% of the entire population. Estimate the fraction of students at your school who are left-handed.

4. A study by Camilla Benbow and Julian Stanley of children younger than 14 who scored 700 or higher on the math SAT found that 20% were left handed, as compared to 8% of the entire population. Estimate the fraction of faculty at your school who are left handed.

5. Ask at least 50 randomly selected US students whether they expect to receive more or less money from Social Security than their parents will receive. Also ask them how much monthly benefits they expect to receive (in today's dollars). Discarding those who say they will receive the same amount as their parents, estimate the overall percentage of students who believe they will receive more than their parents. Also estimate the average monthly benefit that students expect to receive.

6. College students are said to experience the Frosh 15—an average weight gain of 15 pounds during their first year at college. Test this folklore by asking at least 100 randomly selected second-year students how much weight they gained or lost during their first year at college. Estimate a 95% confidence interval for the population mean.

7. People experiencing an earthquake often grossly overestimate how long the quake lasted; for example reporting that a 6-second quake lasted 30 seconds. Show a random sample of students a memorable event, such a snippet of loud music or you dancing, and then ask them how long this event lasted. Do your data indicate that people are more likely to underestimate or overestimate how long the event lasted?

8. Persi Diaconis, a famous mathematician, claims that a spun US penny will land tails about 80% of the time. Put a clean Lincoln penny on a flat surface and hold the coin on its edge with one finger. Then flick the coin with a finger on your other hand. Repeat this experiment 100 times and estimate a 95% confidence interval for the probability that a spun penny will land tails.

9. Tell someone that you randomly selected a MLB player who happened to have gotten 174 base hits in 612 times at bat, a .284 batting average. Ask them to estimate a range for the probability that this player will get a base hit in his next time at bat, such that they are

95% sure that the true probability of this player getting a base hit is inside this range. Compare this to the actual 95% confidence interval, assuming that batting can be described by the binomial distribution.

10. Assemble a very large number of marbles, slips of paper, or other objects that are of two different colors or other features; for example, a very large number of marbles, of which 70% are red and 30% are blue. Place the marbles inside a closed bag or box and allow someone to reach into the bag or box and withdraw ten marbles and then, based on their results, estimate a range for the fraction of all marbles that are red (or blue), such that they are 95% sure that the true percentage is inside this range. Use this calculator <https://statpages.info/confint.html> to determine the exact 95% confidence interval. Repeat this experiment at least 40 times and see how often the ranges by the participants include the true percentage.

11. Show someone a large transparent jar of pebbles, jelly beans, or other small objects and ask them to estimate a range for the number of objects in the jar, such that they are 95% sure that the actual number of objects is in this range. Repeat this experiment at least 40 times. How often did the specified ranges include the actual number of objects?

12. Show someone a large transparent jar of pebbles, jelly beans, or other small objects and ask them to estimate the number of objects in the jar. Repeat this experiment at least 50 times. Does a 95% confidence interval include the actual number of objects? How many individuals made guesses that were more accurate than the average of all the guesses?

13. Have ten or more friends stand in a line outside a closed door, with you the last person in line. Tell your friends standing in line that if anyone asks why they are waiting in line, they should say, "I don't know. I just saw this line and figured it must be something good." Estimate the probability that a person given this answer will join the line.

One-Sample Tests

1. Find someone who claims to have extrasensory perception (ESP) and test this claim.

2. Morningstar and other Internet websites have data on the total returns (dividends plus capital gains) for mutual funds in various fund categories (e.g., US Equity Small Value). Pick a category within US Equity Funds and select a horizon (e.g., 5 years). Now determine

the p-value for a test of the null hypothesis that these data were drawn from a population with a mean equal to the return on the S&P 500 index over this horizon. (Make sure that your S&P 500 return is not just the change in the S&P, but the S&P total return, dividends plus capital gains.) Also use these data to determine the p-value for a test of the null hypothesis that a randomly selected mutual fund has a .5 probability of doing better than the S&P 500 index.

3. A high school basketball coach said that a missed free throw by a right-handed shooter is more likely to bounce to the right, while the reverse is true of a left-hander. To investigate this claim, find five avid basketball players and ask each of them to shoot 100 free throws. Do not tell them the purpose of this experiment, which is to determine if a missed free throw is equally likely to bounce to the same or opposite side as their shooting hand. Use your data to calculate the two-sided p-value for testing the null hypothesis that missed free throws are equally likely to bounce to either side.

4. Young children who play ice hockey are separated by age. In 2024, for example, children born in 2017 were placed in the 7-year-old league and children born in 2018 were placed in the 8-year-old league. A student with a December 11 birthday observed that children with birth dates early in the year are months older than those with later birth dates—someone born in January 2018 is eleven months older than someone born in December 2018. Because coaches give more attention and playing time to better players, this student suspected that children with early birth dates have an advantage when they are young that might cumulate over the years. To test this theory, he looked at the birth dates of 1,487 National Hockey League players and found that 934 had birth dates during the first 6 months of the year. Test his theory using data for another group of professional athletes; for example, male or female soccer players.

5. Calculate the percentage change in the Dow Jones Industrial Average from the close on Thursday the 12th to the close on Friday the 13th for every Friday the 13th beginning in 1980. Is the average percentage change substantial? Determine the two-sided p-value for a test of the null hypothesis that the mean percentage change is 0.

6. The Santa Claus Rally claims that the stock market does unusually well the week before Christmas, December 18 through December 24. A variation is that the market does unusually well the week after Christmas, December 26 through January 1. Test each of these two theories with data on the daily percentage changes in the S&P 500 during at least the past 20 years.

7. Use a single newspaper or national newsmagazine to collect pictures of the winner of a presidential election, some printed a month before the election and an equal number printed a month after the election.

All the pictures should be from the same newspaper or magazine and be full face shots of approximately equal sizes. Do not otherwise screen the photos for being attractive or unattractive. Ask a random sample of students to pick the photo that they consider to be the most flattering and the photo that they consider to be the least flattering. Are their choices equally likely to be from the pre-election and post-election categories?

8. Select two prominent newspapers, magazines, or news websites with very different political reputations (e.g., CNN and Fox). Collect an equal number of pictures from each source of a presidential candidate printed a month before the election. All of the pictures should be full face shots of approximately equal sizes. Do not otherwise screen the photos for being attractive or unattractive. Ask a random sample of students to pick the photo that they consider to be the most flattering and the photo that they consider to be the least flattering. Are their most-flattering choices equally likely to be from each source? What about their least-flattering choices?

9. Conduct a taste test of either Coke versus Pepsi or Diet Coke versus Diet Pepsi. Survey at least 50 students who identify themselves beforehand as cola drinkers with a definite preference for one of the brands you are testing. Calculate the fraction of your sample whose choice in the taste test matches the brand identified beforehand as their favorite. (Do not tell your subjects that this is a test of their ability to identify their favorite brand; tell them it is a test of which cola tastes better.) Calculate the p-value for a test of the null hypothesis that there is a .5 probability that a cola drinker will choose his or her favorite brand.

10. Do more expensive cookies taste better than less expensive ones? Choose two brands of cookies that are similar in appearance but cost quite different amounts. Ask at least 50 persons to taste an unlabeled cookie from each brand and to rate each cookie on a scale of 1–10. Calculate the difference in each person's score and see if the average difference is substantial and statistically persuasive.

11. The *pink tax* says that clothing marketed as female clothing tends to cost more than identical clothing marketed as male clothing. Test this theory by going to a store (online is okay) that sells seemingly identical male and female clothing. Record the price difference for each item and see if the average difference is substantial and statistically persuasive.

12. The *blue tax* says that athletic clothing marketed as male clothing tends to cost more than athletic clothing marketed as female clothing. Test this theory by going to a store (online is okay) that sells seemingly identical male and female athletic clothing. Record the

price difference for each item and see if the average difference is substantial and statistically persuasive.

13. The *pink tax* says that products marketed as female products tend to cost more than identical products marketed as male products. Test this theory by seeing whether women's shaving cream tends to be more or less expensive than men's shaving cream. Go to a store that sells seemingly identical male and female shaving cream made by the same company. Record the price difference for each shaving cream and see if the average difference is substantial and statistically persuasive. (You can use a different product if you want.)

14. Ask a random sample of 50 students to pick a sibling or friend who is the same sex as them. Then ask questions such as these: "Who do you think is more likely to suffer from food poisoning this year, you or [this other person]?" "Who is a better driver, you or [this other person?]" "Who is likely to live longer?" Test the null hypothesis that people are equally likely to say themselves or someone else.

15. Ask 100 randomly selected students to tell you whether they were the first-born (or only) child in their family. Calculate the p-value for a test of the null hypothesis that the probability that a student at your school is first-born (or only) is equal to the national percentage of 40%.

16. Some people slip shoes that have shoelaces on and off without untying the shoelaces. Of those who do untie their shoelaces, some people untie the laces before taking their shoes off; others slip off their still-tied shoes and untie the laces later, when they are putting their shoes back on. What percent of the students who untie their shoelaces, either before taking them off or before putting them back on, do you think would answer "before" to this question: "When you take off shoes that have shoe laces, do you generally untie the laces before taking your shoes off or before putting them back on?" Survey 100 randomly selected students and test your prediction.

17. Do a matched-pair comparison of the calories in McDonald's menu items in two different countries; for example, the United States and the United Kingdom. Is the average difference substantial and statistically persuasive?

18. There are 20 clubs in the English Premier League (EPL). During the course of a season, each club plays each of the other 19 clubs twice, once on its home field and once away, on the opposing team's home field. There are a total of 190 unique two-team pairings. For each of these 190 games, calculate the differential between the score difference for the home game and for the away game. For example, in the 2021 season, Manchester City defeated Manchester United by a score of 4-1 in the game played at Manchester City's home field and

Manchester City defeated Manchester United by a score of 2-0 in the game played at Manchester United's home field. The differential between the home difference and the away difference is 1, regardless of whether we look at it from Man City's viewpoint, $(4-1) - (2-0) = 1$, or Man United's viewpoint, $(0 - 2) - (1 - 4) = 1$. Calculate the home-minus-away differential for all 190 pairings in a recent season and test the null hypothesis that the average differential is zero.

19. Each team in the English Premier League (EPL) plays each of the other 19 teams twice every season—once on its home field and once away, on the opposing team's home field. Choose a recent season and determine the number of red and yellow cards that were given to each team in each game. Test the null hypothesis that the expected value of the difference between the number of cards given to a team does not depend on whether the game is played at home or away.

20. Persi Diaconis, a famous mathematician, claims that a spun U.S. penny will land tails about 80% of the time. Put a clean Lincoln penny on a flat surface and hold the coin on its edge with one finger. Then flick the coin with a finger on your other hand. Repeat this experiment 100 times and test the null hypothesis that the probability that a spun penny will land tails is .80.

21. Choose an essay that a student has written for a school class. Ask ChatGPT or another large language model (LLM) to write an essay on the same subject. Show the two essays to a random sample of 30 students and ask each student to identify the LLM-generated essay. Use your results to test the null hypothesis that each essay is equally likely to be chosen.

Two-Sample Tests

1. What percentage of college seniors expect to be married within 5 years of graduation? What percentage expect to have children within 5 years of graduation? How many biological children do college seniors expect to have during their lifetimes? Are the differences between the male and female responses substantial and statistically persuasive?

2. Do economics majors get better grades in humanities courses or in science courses? (Do not include economics courses.) Is the average difference substantial and statistically persuasive?

3. Ask 50 males and 50 females this question: "You were engaged to be married to A, but broke off the engagement a month before the

wedding when you found A in bed with your best friend. Two years have now passed and you are engaged to be married to someone else. A month before the wedding, A calls you and says 'Let's get drunk tonight for old-time's sake.' Do you say yes or no?" Devise a procedure that will allow each person to respond anonymously, but will still allow you to distinguish male and female responses (e.g., different colored sheets of paper). Are the differences in the male and female responses substantial and statistically persuasive?

4. Ask 50 males and 50 females to write down a brief answer to this question: "You are engaged to be married to A and while you are both at a party, a former lover who is unknown to A flirts with you outrageously. You rebuff these advances, but afterwards A asks you to identify the person who was flirting with you. What do you say?" Devise a procedure that will allow each person to respond anonymously, but will still allow you to distinguish male and female responses. Now ask someone who is unaware of your system for identifying male and female responses to classify each response as either truthful or deceitful. Are the differences in the male and female responses substantial and statistically persuasive?

5. In the 1980s Kahneman and Tversky asked students the following question: "Imagine that you have decided to see a play and paid the admission price of $10 per ticket. As you enter the theater, you discover that you have lost the ticket. The seat was not marked and the ticket cannot be recovered. Would you pay $10 for another ticket?" They asked a different group of students this question: "Imagine that you have decided to see a play where admission is $10 per ticket. As you enter the theater, you discover that you have lost a $10 bill. Would you still pay $10 for a ticket for the play?" Do this experiment with an updated ticket price and see if there is a substantial and statistically persuasive difference in the responses to these two questions.

6. Ask a random sample of students the following question: "Your favorite singer is performing tonight and you paid $200 for a ticket to the concert. Unfortunately, you will have to drive 60 miles through a snowstorm to get to the concert. Will you go?" Ask another random sample of students this question: "Your favorite singer is performing tonight and you were given a free ticket to the concert. Unfortunately, you will have to drive 60 miles through a snowstorm to get to the concert. Will you go?" Is there a substantial and statistically persuasive difference in the responses to these two questions?

7. Ask a random sample of students the following question: "Your favorite singer is performing tonight and a friend paid $200 for a ticket to the concert. Your friend is sick and gave you the ticket for free. Unfortunately, you will have to drive 60 miles through a

snowstorm to get to the concert. Will you go?" Ask another random sample of students this question: "Your favorite singer is performing tonight and a friend got a free ticket to the concert. Your friend is sick and gave you the ticket. Unfortunately, you will have to drive 60 miles through a snowstorm to get to the concert. Will you go?" Is there a substantial and statistically persuasive difference in the responses to these two questions?

8. A stock's dividend yield is its annual dividend divided by its current market price; its earnings yield is the annual earnings divided by the current price. Determine the dividend and earnings yields on some date in the past for each of the 30 stocks in the Dow Jones Industrial Average. Identify the ten stocks with the highest dividend yields and the ten with the lowest dividend yields. Calculate the percentage price increases for these 20 stocks over the next year. Redo these calculations, this time using the earnings yields. In each case, are the observed differences between the two groups substantial and statistically persuasive?

9. Post a sign on the main entrance to a campus building requesting the use of a less convenient entrance; for example, "Please use the door on the north side of the building." From an inconspicuous location, observe how many people ignore the sign and use the main entrance and how many people do not use the main entrance. Is there a substantial and statistically persuasive difference in the behavior of students and professors? In the behavior of male and female students? Try to pick a building and time when traffic is light, so that large numbers do not enter simultaneously.

10. Choose a random sample of at least 20 barber shops or beauty salons that cut both men's and women's hair. Do not choose more than one store from a chain. Have a female telephone each store in your sample and find the price of the least expensive female haircut. A few hours later, have a male telephone this store and determine the price of the least expensive male haircut. Are the observed differences substantial and statistically persuasive?

11. Anchoring is a general human tendency to rely on a reference point when making decisions. For example, a student did a term paper in which randomly selected students were asked one of these two questions:

> *The population of Bolivia is 5 million. Estimate the population of Bulgaria.*

> *The population of Bolivia is 15 million. Estimate the population of Bulgaria.*

Those who were told that Bolivia's population was 15 million tended to give higher answers than did those told that Bolivia's population was 5 million. Several similar questions confirmed this pattern. People use a known "fact" as an anchor for their guess. Redo this study using questions you make up.

12. Follow the instructions for the preceding project but use an unrelated "fact;" for example,

> *Annual per capita income in Bulgaria is $4,000. Estimate the population of Bulgaria.*

> *Annual per capita income in Bulgaria is $20,000. Estimate the population of Bulgaria.*

13. Ask at least 50 students to participate in this experiment. Tell each volunteer that you are going to ask two brief questions. Then ask the first question: "Guess the number that you will pull from a bag containing 100 slips of paper, numbered 1 to 100." (In fact, 50 slips have 10 written on them; the other 50 slips have the number 65.) After the number is revealed, ask the second question: "Estimate the percentage of total United Nations membership made up of countries in Africa." Test whether there is a substantial and statistically persuasive difference between the Africa guesses made by those who picked 10 and those who selected 65.

14. Arrange 16 facial photos in a 4 × 4 grid, with 3 of the faces smiling and the other 13 faces neutral. Ask a random sample of people to pick out the three smiling faces and record how long it takes each person to do so. Repeat the same experiment with three angry faces and 13 neutral faces. Does it take longer, on average, for people to recognize smiling faces or angry faces?

15. Ask someone to shoot ten basketball free throws, then ask them to shoot two free throws blind-folded, and then ten more free throws not blindfolded. Repeat this experiment with several different people, but don't let anyone in your sample see your experiment with other people. For half the people in your sample, loudly cheer after each blindfolded shot and tell the person that they made the shot; for the other half, groan after each blindfolded shot and say they missed badly. Does the negative/positive reinforcement during the blindfolded shots have any effect on their performance when not blindfolded?

16. Ask 50 female students these four questions: "Among female students at this school, is your height above average or below average?" "Is your weight above average or below average?" "Is your intelligence above average or below average?" "Is your physical attractiveness above average or below average?" Ask 50 male students these same questions (in comparison to male students). Try to design a survey procedure that will ensure candid answers. For each sex and each question, test the null hypothesis that $p = .5$. Also, compare the male and female answers to each question.

17. Ask 50 male and 50 female students these two questions: "Compared to all male students nationwide, is the intelligence of the average male student at your school above average or below average?" "Compared to all male students nationwide, is the physical attractiveness of the average male student at your school above average or below average?" Repeat the survey with a new sample, asking 50 male and 50 female students these two questions: "Compared to all female students nationwide, is the intelligence of the average female student at your school above average or below average?" "Compared to all female students nationwide, is the physical attractiveness of the average female student at your school above average or below average?" Try to design a survey procedure that will ensure candid answers. For each survey question, are there substantial and statistically persuasive differences in the male and female answers?

18. Stand in a place where a reasonable number of people walk by. Stare at the sky and have a friend record the percentage of the people walking by who look up at the sky, too, and the percentage who do not. Now repeat this experiment at the same place with 5, 10, or 15 of your friends staring at the sky. Again, have a friend record the percentage of people walking by who look up at the sky and the number who do not. Is the difference substantial and statistically persuasive? (Choose a location where people do not walk by in large groups.)

19. Have ten or more friends stand in a line outside a closed door, with you stationed as the last person in line. Tell your friends standing in line that if anyone asks why they are waiting in line, they should say, "I don't know. I just saw this line and figured it must be something good." Is there a substantial and statistically persuasive difference in the propensity of male and female students to join the line?

20. Ask at least 20 people this question: "A bag contains 10 red balls and 10 blue balls. You randomly pick a ball from the bag and win $10 if the ball is blue. What is the maximum amount you would pay to play this game once?" Ask another group of at least 20 people this question: "There are an unknown number of red and blue balls in a bag. You pick a color, red or blue, and then randomly pick a ball from the bag and win $10 if the ball is the color you picked. What is the maximum amount you would pay to play this game once?" Is there a substantial and statistically persuasive difference in the responses to these two questions?

21. Find a crosswalk at an intersection with traffic lights that has a walk button that pedestrians can use to signal that they are waiting to cross the street. Record several observations on how long it takes the light to turn green after having turned red: (a) when no one pushes the pedestrian walk button; and (b) when you push the pedestrian

walk button shortly after the light turns red. Is the average difference substantial and statistically persuasive?

22. Select two NCAA basketball tournaments—one recent and the other several years ago—and tabulate the number of three-point shots and the total number of shots taken by each team in each game. Compare the fraction of the shots taken that were three-point shots (a) in total in each tournament; (b) by the favorite and underdog teams in the recent tournament; and (c) by the favorite and underdog teams in the distant tournament. Are these various differences substantial and statistically persuasive?

Chi-Square Tests

1. Set up a mock ESP experiment by writing the numbers 1 through 10 on ten identical pieces of paper, and placing these in a hat, bag, or other opaque container. Now tell a randomly selected person that you are going to select one of these pieces of paper and concentrate on the number, while the subject tries to read your mind. Be very careful to ensure that the subject cannot see the number on the paper. Record the answer and then tell the subject the selected number. Repeat this experiment with at least 50 different subjects. When you have all of your data, calculate the p-value for a test of the null hypothesis that each number is equally likely to be chosen by the subjects.

2. After attempting to imitate a popular television character, a young man concluded that whether one is right-handed or left-handed affects how far apart the two middle fingers on each hand can be spread. If a wider "V" is made with the right hand, the person is probably left-handed; if a wider V is made with the left hand or there is no difference, the person is probably right-handed. Ask at least 100 randomly selected students to spread the middle two fingers on each hand to make a "V." Record whether the wider "V" is made with the right hand or the left hand (or no difference), and then ask the person whether he or she is left-handed or right-handed. How well does the wider "V" predict handedness?

3. Ask at least 100 students, "Do you usually make sure you look good before leaving your room?" Also record the person's sex and year in school. Determine the p-values for a test of these null hypotheses: (a) the responses are unrelated to sex, and (b) the responses are unrelated to year in school.

4. Ask randomly selected students if they have had a serious romantic relationship in the past 2 years and, if so, to identify the month in

which the most recent relationship began. When you have found 120 students who answered yes and identified the month, calculate the p-value for a test of the null hypothesis that each month is equally likely for the beginning of a romantic relationship.

5. Ask randomly selected students if they have ended a serious romantic relationship in the past 2 years and, if so, to identify the month in which the most recent ended-relationship ended. When you have 120 students who answered yes and identified the month, calculate the p-value for a test of the null hypothesis that each month is equally likely for the end of a romantic relationship.

6. In the game Roshambo (rock-scissors-paper), two players simultaneously move their fists up and down three times and then show a fist (rock), two fingers (scissors), or an open hand (paper). Rock beats scissors, scissors beats paper, and paper beats rock. Play this game against at least 120 different people, recording the initial move of each opponent. Use these data to test the null hypothesis that rock, scissors, and paper are used equally often on the initial move. Do your results support the adage, "Losers lead with rock," based on the perception that naive players lead with rock more than one-third of the time?

7. Ask 100 randomly selected students this question and then compare the male and female responses: "You have a coach ticket for a nonstop flight from Los Angeles to New York. Because the flight is overbooked, randomly selected passengers will be allowed to sit in open first-class seats. You are the first person selected. Would you rather sit next to: (a) Bill Clinton; (b) Hillary Clinton; or (c) Michael Jordan? Alternatively, you can choose three other famous people.

8. Use the obituaries in a book of famous people, *The New York Times*, or another source to find the birth and death dates of at least 120 persons. Divide these data into four categories: deaths that occurred: during the 14 days preceding the birthday, on the birthday, during the 14 days following the birthday, and on other days. Calculate the p-value for a test of the null hypothesis that a person's death date is not related to his or her birth date.

9. Make a list of ten well-known books (including one that you feel will be controversial) and ask at least 30 teachers and 30 students to separate these books into two groups of five, based on how important it is that students read these books: most important and least important. Are there statistically persuasive differences between how the students and teachers classified the book you felt would be controversial?

10. The nine positions on a baseball team can be divided into four categories: pitcher, catcher, the four infielders, and the three outfielders. Collect all the data you can on MLB managers and test the null

hypothesis that, among those managers who played baseball, the probabilities of having played in these four categories are 1/9, 1/9, 4/9, and 3/9, respectively.

11. Follow the instructions for the preceding exercise but for a different sport such as soccer, football, or basketball.

12. Administer the following four tests to at least 50 subjects, and then apply a chi-square test with the columns right-handed or left-handed and the rows tests a, b, and c.

   a. Ask the subject to stand with his or her back to you. Then ask the subject to jump around in a single motion to face you. Record whether the person jumps clockwise (pushing off with a dominant left foot) or counterclockwise (pushing off with a dominant right foot).

   b. Ask the subject to look at an object 10 feet away through a tube made with the hands held a foot in front of his or her face. Close or cover first one eye and then the other and record whether the subject can still see the object through the tube when the left eye is open (left-eye dominance) or when the right eye is open (right-eye dominance).

   c. Ask the subject to put his or her hands together behind the head, with the fingers interlaced. Record whether the thumb on the bottom (the dominant thumb) is from the left or right hand.

   d. Ask the subject whether he or she is left-handed or right-handed.

13. Ask a random sample of 100 students to compare themselves to other students in the population the sample was taken from. For example, if you survey a random sample of ABC School students, ask each student to compare himself or herself to other ABC students. Ask the students to rank themselves in their ability to get along with others; top 25%, next 25%, next 25%, bottom 25%. What is the p-value for a test of the null hypothesis that a randomly selected person is equally likely to pick each of these four quartiles?

14. Ask 100 randomly selected students to tell you their birth order: only child, first-born, second-born, third-born, or other, and compare these to the national percentages.

Simple Regression

1. For the two most recently completed major league baseball seasons, identify those players with at least 502 official times at bat in each season. Use the simple regression model to see how well each

player's batting average in the most recent season is predicted by his batting average the preceding season. Do these results exhibit regression toward the mean?

2. Imagine that you are a US government statistician in the 1860s. Use the census data for the years 1790–1860 to predict the US population in 1930. To do so, estimate the equation $Y = \alpha + \beta X + \varepsilon$, where Y is the population and X is the year (1790, 1800, and so on). Repeat with Y equal to the natural logarithm of population and compare your results.

3. Has there been any long-term trend in voter turnout in US presidential elections? Does the Democratic or Republican presidential candidate tend to do better when there is a heavy turnout?

4. Select a US President who ran for office twice; for example, Bill Clinton, George Bush, Barack Obama, or Donald Trump. For each of the 50 states, calculate this person's percentage of the total votes cast for the Democratic and Republican presidential candidates in each year; do not include the votes cast for other candidates. Is there a statistical relationship between these two sets of data? Are there any apparent outliers or anomalies?

5. An old Wall Street saying is, "As January goes, so goes the year." Use the simple regression model to see whether the February-through-December percentage change in the Dow Jones industrial average of stock prices is well predicted by the percentage change in January.

6. Use the simple regression model to see whether the annual percentage-point change in the interest rate on 10-year Treasury bonds is well predicted by the change the preceding year.

7. Use the simple regression model to see whether the annual percentage change in the Dow Jones industrial average of stock prices is well predicted by the percentage change the preceding year.

8. It has been argued that interest rates move up and down in lockstep with the rate of inflation. Test this theory by using annual data to estimate the model, $R = \alpha + \beta P + \varepsilon$, where R is the interest rate on 1-year Treasury bonds and P is the annual rate of inflation.

9. Are future interest rates better predicted by current interest rates or by the current rate of inflation? Use annual data to estimate these two models, $R = \alpha_1 + \beta_1 S + \varepsilon_1$, and $R = \alpha_2 + \beta_2 P + \varepsilon_2$, where R is the interest rate on 1-year Treasury bonds, S is interest rate the previous year, and P is the annual rate of inflation the previous year.

10. Find the heights and weights of the quarterbacks, running backs, and wide receivers who have been inducted into professional football's hall of fame. For each of these three positions, estimate a linear regression model with height as the dependent variable and time as

the explanatory variable and another linear regression model with weight as the dependent variable and time as the explanatory variable. In each case, let the time variable equal the year in which the player first played professional football.

11. Pick a date and approximate time of day (e.g., 10:00 in the morning on April 1) for booking nonstop flights from an airport near you to at least a dozen large US cities. Determine the cost of a coach seat on each of these flights and the distance covered by each flight. Use your data to estimate a simple regression model with ticket cost the dependent variable and distance the explanatory variable. Are there any outliers?

12. Collect data for at least 50 years on the cost of attending your school (or another school that interests you). Taking into account the increase in the overall price level during these years, use the simple regression model to see whether there has been a trend in the real cost of attending this school.

13. Investigate how well mutual fund performance is predicted by past performance.

14. Use data for several presidential elections to estimate a simple regression model, where Y = percent of two-party vote for US President received by the incumbent party's candidate and X = percent change in real disposable income in the last year of the incumbent party's term. Is there a statistically persuasive and substantial effect? Which elections appear to be outliers or anomalies?

15. This website <https://www.spotrac.com/mlb/rankings/_/year/2024> has MLB player salaries. This site < https://www.espn.com/mlb/war/leaders/_/type/seasonal/year/2024 > has data for player wins above replacement. Looking at the 30-50 highest-paid players for the most recently completed season, how strong is the relationship between pay and performance?

16. Looking at the most recently completed MLB season, how strong is the relationship between team payroll and regular-season team performance?

17. Looking at the most recently completed National Basketball Association (NBA) season, how strong is the relationship between team payroll and regular-season team performance?

18. Looking at the most recently completed English Premier League (EPL) season, how strong is the relationship between club payroll and regular-season club performance?

19. This website <https://www.multpl.com/shiller-pe/table/by-year> has data on the Shiller PE ratio (CAPE) on January 1 of every year going back to 1872. This website <https://www.multpl.com/s-p-

500-historical-prices/table/by-year > has data on the S&P 500 on January 1 of every year going back to 1871. Starting with January 1, 1872, how well does the Shiller P/E ratio at the start of each year predict the percentage change in the S&P 500 that year?

20. Ask a random sample of US citizens to think of the last two digits of their Social Security number. Then show them a picture of a bottle of wine (or some other item of uncertain value) and ask them to guess the price of the wine. Do people with higher Social Security numbers tend to guess higher prices?

21. One measure of the relative importance of skill and luck in competitions is the consistency of the outcomes. Calculate the correlation between the winning percentages of Major League Baseball (MLB) teams before and after the All-Star Break. Compare this to the correlation for National Basketball Association (NBA) teams before and after their All-Star Break.

22. Follow the instructions for the previous question, comparing the correlations for National Football League (NFL) teams and English Premier League (EPL) teams for the first-half and second-half of their seasons. You can use winning percentages for NFL teams but may want to use points for EPL teams since many games are drawn.

Multiple Regression

1. Ask 100 randomly selected students to write down their sex, grade point average, birth order (first-born or only child, middle-born, or last-born), and height. Now estimate a multiple regression model with grade point average as the dependent variable and sex, birth order, and height as the explanatory variables.

2. Construct a plausible multiple regression model for predicting the first-year salaries of college seniors who are looking for jobs. Now use a random survey to gather data on their predicted salary (or actual salary if they already have a job) and your explanatory variables, and use least squares to estimate your model.

3. Use RealEstate.com, ZipRealty.com, Zillow.com, or a similar website to find the prices for houses that are for sale in your hometown or where you plan to live in the future. Specify a plausible multiple regression model for explaining these prices and use a random sample of at least 60 houses to estimate your model's parameters.

4. Select a car model and year (e.g., 2020 Camry), and estimate a plausible multiple regression model for predicting used car prices.

5. Use data for several presidential elections to estimate a multiple regression model, where Y = percent of the two-party vote for US President received by the incumbent party's candidate, X1 = percent of the two-party vote for US President received by the incumbent party's candidate in the previous presidential election, and X2 = percent change in real disposable income in the last year of the incumbent party's term. In the 2020 election, for example, Donald Trump (a Republican) was the incumbent, so Y is the percent of the 2020 two-party vote received by the Republican candidate; X1 is the percent of the 2016 two-party vote received by the Republican candidate; and X2 is the percent change in real disposable income in 2020. Are the coefficients of both explanatory variables statistically persuasive and substantial? Which elections appear to be outliers or anomalies?

6. Use data for each of the 50 states and one presidential election to estimate a multiple regression model, where Y = percent of the two-party vote for US President received by the incumbent party's candidate in this state in this election year, X1 = percent of the two-party vote for US President received by the incumbent party's candidate in this state in the previous presidential election, and X2 = percent change in the unemployment rate in this state in the last year of the incumbent party's term. In the 2020 election, for example, Donald Trump (a Republican) was the incumbent, so Y for Iowa in 2020 is the percent of the 2020 two-party vote in Iowa received by the Republican candidate; X1 is the percent of the 2016 two-party vote in Iowa received by the Republican candidate; and X2 is the percent change in the unemployment rate in Iowa in 2020. Are the coefficients of both explanatory variables statistically persuasive and substantial? Which states appear to be outliers or anomalies?

7. Collect daily data during two recent years for Y = the percentage price change in the S&P 500 and for 100 explanatory variables, each equal to the daily difference between the high and low temperatures in 100 diverse cities. Use the first year of your data to estimate 100 simple regression models, each using one of your 100 explanatory variables. Identify the five cities with the highest t-values and estimate a multiple regression model using only these five cities as explanatory variables. Finally use the estimated coefficients for the multiple regression model you estimated with the first year's data to predict the daily percentage price changes in the S&P 500 during the second year. What is your conclusion?

8. Collect daily data during two recent years for Y = the percentage price change in the S&P 500 and for 100 explanatory variables, with the daily value of each explanatory variable equal to a randomly generated number. Use the first year of your data to estimate 100 simple regression models, each using one of your 100 explanatory

variables. Identify the five variables with the highest t-values and estimate a multiple regression model using only these five explanatory variables. Finally use the estimated coefficients for the multiple regression model you estimated with the first year's data to predict the daily percentage price changes in the S&P 500 during the second year. What is your conclusion?

9. Use a random number generator to create 500 observations for 101 variables. Let the first random variable be Y, the variable you want to predict and let the other 100 variables be the potential explanatory variables that will be used to predict Y. Use the first 250 observations to estimate a multiple regression model using the 100 explanatory variables to predict Y. (Alternatively, you can estimate 5 models with 20 explanatory variables.) Identify the five explanatory variables with the highest t-values. Reestimate your multiple regression model using these 5 explanatory variables and the first 250 observations to predict Y.

Now, use the estimated coefficients for the second multiple regression model that you estimated with the first 250 observations to predict the values of Y for the second 250 observations. What is your conclusion?

10. Select a 10-year period, such as 2015–2024. Estimate a multiple regression model using the following four explanatory variables, all recorded the year before the Super Bowl, to predict the Super Bowl score differential, AFC team score minus NFC team score, in the first 5 years of your data:

High temperature in London on June 30

Dow Jones Industrial Average on June 30

Average number of letters in the names of Physics Nobel Prize winner(s) that year

Number of points earned by the soccer team winning the Premier League in the season ending that May

How well does your model estimated from the first 5 years predict the Super Bowl score differential in last 5 years of your data? What is your conclusion?

11. Use a multiple regression model with the high temperatures on election day in these ten cities as explanatory variables for predicting the Democratic Party candidate's percentage of the total two-party vote for President in the 11 presidential elections 1980–2020: Claremont, California; Bozeman, Montana; Broken Bow, Nebraska; Burlington, Vermont; Caribou, Maine; Cody, Wyoming; Dover, Delaware; Elkins, West Virginia; Fargo, North Dakota; and Pocatello, Idaho. Now use your estimated model to predict the Democratic Party candidate's

percentage of the total two-party vote for President in 2024. What is your conclusion?

12. This website <https://www.spotrac.com/epl/payroll/> has data on English Premier League (EPL) club payrolls separated into forwards, midfielders, defensemen, and goalkeepers. Looking at the most recently completed season, how well is club performance predicted by the payroll in these four categories?

13. This website <https://www.premierleague.com/stats/top/clubs/wins?se=489> has data on English Premier League (EPL) club wins, red cards, and yellow cards. Looking at the most recently completed season, how well is club performance predicted by the number of red cards and yellow cards?

14. Are future interest rates better predicted by current interest rates or by the current rate of inflation? Use annual data to estimate the model, $R = \alpha + \beta_1 S + \beta_2 P + \varepsilon$, where R is the interest rate on one-year Treasury bonds, S is interest rate the previous year, and P is the annual rate of inflation the previous year.

15. This website <https://www.multpl.com/s-p-500-historical-prices/table/by-year > has data on the S&P 500 on January 1 of every year going back to 1871. This website <https://www.multpl.com/s-p-500-earnings-yield > has data on the S&P 500 earnings yield on January 1 of every year going back to 1871. This website <https://www.multpl.com/s-p-500-dividend-yield/table/by-year> has data on the S&P 500 dividend yield on December 31 of every year going back to 1871. How well is the percentage change in the S&P 500 each year predicted by the earnings yield and dividend yield at the start of that year?

16. Create and estimate a model for predicting the number of medals a country wins at the Summer Olympics. You might include a nation's population and per capita income as explanatory variables. Add other variables that you think are important. Identify nations that are under-achievers and over-achievers according to your model.

Index

A

A/B test, 148
ANOVA F-statistic tests, 244,
    260–263
Approximation, 101, 120, 123, 125,
    127, 131, 136–139, 143,
    145–146, 243
Asymmetrical distribution, 7
Average, 1–10, 123
    distribution, 7
    number, 170

B

Bar chart, 25, 35
Bayes' rule, 78–92, 95–97, 265
Bayesian analysis, 88, 92, 202, 237
Bayesian posterior probability, 78
Bernoulli trial model, 83, 101, 104–106,
    110–111, 260
Bertrand's box paradox, 96
Bimodal distribution, 128
Binomial distribution, 99–112, 127–128,
    139, 143–144, 243

C

Central limit theorem, 127, 140–141, 155,
    166, 209, 243–244
Chi-square
    distribution, 181
    statistic, 178, 180–181, 184, 186,
        260–264
    tests, 167, 170–188, 280–282
Confidence intervals (CI), 131–132, 134,
    140, 143, 269–271
    and one-sample tests, 130–147
    and two-sample tests, 148–169
Contingency table, 87–89, 95, 97
Correlation, 31–32, 50, 168, 189–190,
    193, 196–198, 201, 203, 209,
    227, 241–242, 245, 253, 257,
    263–264, 285

D

Data, misleading, 36–51
Degrees of freedom, 160, 164, 177,
    180–183, 185–186
Descriptive statistics, 1–10, 222, 255, 266
Difference-in-means
    formula, 162
    t-test, 245–248, 252
    test, 150, 160–162, 165, 167–169, 236,
        243
Difference-in-proportions test, 159, 161,
    163, 165, 167, 242, 257, 260–262

G

Gambler's fallacy, 213
Graphs, 11–26, 266
    horizontal axis, 27, 31–32
    vertical axis, 27–28, 31, 34

H

HARKing, 257–258, 263–265
Histograms, 18, 22, 25–26, 34–35, 119,
    121, 125, 127, 242, 244, 266, 268
Home-field advantage, 3, 8

L

Law of averages, 113–118
Least squares regression, 190–191, 195,
    199, 202

M

Margin of error, 130, 139
Matched-pair test, 149–151, 153,
    160–165
Mean, 1–10, 266
Median, 1–10, 14, 34, 103, 110, 121,
    123–124, 128–129, 154, 196, 222,
    226, 242, 244, 247, 253, 259, 266
Monty Hall problems, 93–98

Multicollinearity, 219, 225, 243
Multiple regression, 217–240, 285–288
Multiplication rule, 65, 67–68, 73,
    75–76, 107

N

Normal approximation, 101, 120, 125,
    127, 136–139, 145, 243
Normal distribution, 119–130, 136, 141,
    143, 147, 172, 241, 244, 259–260,
    269
Null hypothesis, 135–136, 138, 141–142,
    144, 152, 155, 158, 162, 165–166,
    171, 173, 176, 179–180, 182, 184,
    186, 188, 200, 202–203, 226, 232,
    236, 258, 265

O

Omitted-variable bias, 258
One-sample tests, 271–275
    and confidence intervals (CI),
        130–147
One-standard-deviation rule of thumb,
    129

P

P-hacking, 258
P-value, 140–141, 149–152, 159–160,
    163–164, 180, 182, 184–185, 202,
    226, 232
Pie chart, 25, 35
Posterior probability, 84
Probabilities, 51–83, 93–96, 98–103,
    105–108, 120, 125, 131, 134, 138,
    142, 146, 152, 161, 171, 181, 202,
    267–269
Probability distribution, 127, 131,
    135, 143
Probability of guilt, 84

R

R^2, 190, 201, 224, 230, 232, 234
Random number generator, 148, 159,
    254, 287

Regression, 196, 202
    equation, 189, 191
    model, 224
    toward the mean, 205–216, 264
Revised probability, 78–80, 83, 86
Run-of-luck theory, 113

S

Scatter diagram, 25–26, 35, 201
Self-selection bias, 45–47, 49–50, 258
Side-by-side box plots, 25–26, 35, 158
Simple regression, 189–205, 282–285
Simpson's paradox, 2, 7–8
St. Petersburg paradox, 67
Standard deviations, 5, 10, 120–121, 131,
    136, 139, 143, 266
    of mean, 268
Standard errors, 189, 192–193, 202,
    217–219, 221, 225–228, 230–232,
    237–238
Statistics-based economic analysis, 121
Survivorship bias, 45–50, 201, 258
Symmetrical distribution, 34

T

t-test, 240
t-value, 144, 152, 154, 160
Thomas Paine's argument, 86
Time series graph, 25–26, 35, 266
Two-sample tests, 275–280
    and confidence intervals (CI), 148–169
Two-sided p-value, 27, 109, 132, 135, 138,
    141, 143–146, 150–153, 159–169,
    177, 182, 200, 218, 231, 233, 260,
    265, 272

W

Wins above replacement (WAR),
    206, 268

Z

Z-test, 244, 260–262
Z-values, 125–129, 138–140, 147, 158–160,
    163, 166